D0146731

Krisciunas, Kevin,
1953-

Astronomical
centers of the
world

$24.95

DATE			

Astronomical centers of the world

Astronomical centers

of the world

KEVIN KRISCIUNAS

Joint Astronomy Centre, Hilo, Hawaii

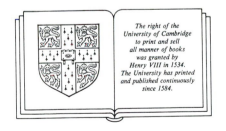

The right of the
University of Cambridge
to print and sell
all manner of books
was granted by
Henry VIII in 1534.
The University has printed
and published continuously
since 1584.

CAMBRIDGE UNIVERSITY PRESS

Cambridge

New York   New Rochelle   Melbourne   Sydney

Published by the Press Syndicate of the University of Cambridge
The Pitt Building, Trumpington Street, Cambridge CB2 1 RP
32 East 57th Street, New York, NY 10022, USA
10 Stamford Road, Oakleigh, Melbourne 3166, Australia

First published 1988

Printed in Great Britain at the University Press, Cambridge

British Library cataloguing in publication data

Krisciunas, Kevin
Astronomical centers of the world.
1. Astronomical observatories
I. Title
522′.1   QB81

Library of Congress cataloguing in publication data

Krisciunas, Kevin, 1953–
Astronomical centers of the world.

Includes index.
1. Astronomical observatories – History.
2. Astronomy – Research – History.   I. Title.
QB81.K75   1987   522.1′09   86-21542
ISBN 0 521 30278 1

TM

Contents

Preface

If the cost of gathering celestial treasure exceeds that of searching for the buried chests of a Morgan or a Flint, the expectation of rich return is surely greater and the route not less attractive. Long before the advent of the telescope, pharaohs and sultans, princes and caliphs built larger and larger observatories, one of them [the Fakhri sextant at Samarkand] said to be comparable in height with the vaults of Santa Sophia. In later times the kings of Spain and of France, of Denmark and of England took their turn, and more recently the initiative seems to have passed chiefly to American leaders in industry.

George Ellery Hale (1928)

The American astronomer B.A. Gould (1824–96), founder of the *Astronomical Journal*, referred to Pulkovo Observatory as the 'astronomical capital of the world'.[1] This was true in the mid-nineteenth century when Pulkovo was producing the most accurate astrometrical measurements. Since the turn of the twentieth century more and more astrophysics has been done compared to positional astronomy. With the establishment of mountaintop observatories closer to the equator, the astronomical capital is certainly somewhere else.

The question naturally arises: Where has the astronomical capital of the world been over the centuries? This book is the result of my selections.

This is not intended as a general history of astronomy, of which there are many excellent ones. The reader will not find detailed discussions in this book of the breakthroughs of such revolutionary scientists as Copernicus, Galileo, Newton or Einstein. The thrust of my study is this: at a given time there was an astronomical center whose equipment and staff were such that crucial, path-breaking work could have been done nowhere else. Or in some cases it could have been done somewhere else, but was not. In general, the personal whereabouts of any individual could not be considered the astronomical capital of the world. The work of Copernicus and Newton for astronomy is without question very important, but both were primarily theoreticians. The astronomy of the last 400 years is a combination of theoretical analyses and observational breakthroughs based on developing technology. This is not to say that I am presenting a history of astronomical technology. Yet astronomical science would not have achieved so much without such masters as Troughton & Simms, Merz & Mahler, or Alvan Clark & Sons. Here I am saluting the observatories and dynamic individuals that were involved in the collaborative progress of institutionalized astronomy. I certainly do not mean to imply that astronomy is more important if it is done at a large institution with a large budget. But the magnificent library at Alexandria, the center of ancient learning, was made possible by royal patronage. Kepler's analysis of Tycho Brahe's observations would not have been possible without the royal Danish funds which supported Tycho's observatory. Tsar Nicholas I provided essentially unlimited funds to Wilhelm Struve for the establishment of Pulkovo Observatory.* Where would nineteenth century positional astronomy have been without that? Would Edwin Hubble have managed to demonstrate the expansion of the universe without the efforts of Hale and the funds of Carnegie?

Before the advent of printing and the invention of the telescope progress in astronomy was sporadic and communication of research was slow. The prime accomplishment of the first fifteen centuries of our era was the preservation of Greek knowledge of the heavens. In spite of the destruction of the Alexandrian Library before the seventh century AD, Greek astronomy was not lost. It was preserved by the Moslems, some of whom were observers as well. Here we shall discuss the contributions of a great fifteenth century (pre-telescopic) observer, Ulugh Beg. For the next astronomical capital of the world we must skip over a century and move to the island observatory at Hven established by Tycho Brahe. The era following Galileo saw the establishment of the Paris (1667) and Greenwich (1675) Observatories. Their existence was firmly rooted in the application of astronomy to navigation, and in both cases national prestige was a strong motivating factor. The relative importance of astrometry was never higher than in the 50 years after the founding of Pulkovo Observatory (1839), during which time no place was more highly regarded. By the end of the nineteenth century the baton of the astronomical capital of the world was passed to America, where it was shared by the Harvard (1843), Lick (1888) and Yerkes (1897) Observatories. Yerkes came about as a result of the efforts of the greatest observatory builder of all time, George Ellery Hale. In 1904 he

---

* An analogous case can be made that makers of history plowed the ground for productive societies and a harvest of scholarly results. Consider: Alexander the Great, the founding of Alexandria, its Museum; Tamerlane the Great, the flourishing of Samarkand, his grandson's observatory; Peter the Great, the founding of St Petersburg and Catherine the Great, the flourishing of St Petersburg, the establishment of Pulkovo Observatory.

also founded Mt Wilson Observatory, and in 1948, 10 years after his death, the 200 inch Hale reflector went into operation at Palomar Mountain.

Today there are many astronomical capitals. Because astronomy has branched out so greatly, where one would hope to do research depends largely on the wavelength of one's observations. If we confine ourselves to the surface of the Earth, the summit of Mauna Kea has no match for infrared and sub-millimeter astronomy. Those who observe at optical wavelengths might prefer the Chilean Andes. The Very Large Array in New Mexico is the most advanced facility for radio astronomy.

Histories of contemporary astronomy such as the *General History of Astronomy* cut off in the middle of this century. There are no general histories of astronomy carried out in the southern hemisphere, and it is too soon to know which greatly heralded observations and associated interpretations of the present will be regarded as correct and significant in twenty or fifty years. It must be left for a future historian to determine for certain where the most crucial observations of the late twentieth century were made and what institutions were responsible for making them happen.

As Will and Ariel Durant point out, only a fool would try to review 2000 years of progress in a single volume. Nevertheless, we proceed.[2] I hope the reader will forgive me for the wide range of source material I have used – from primary sources dealing with Pulkovo Observatory and Mauna Kea, to encyclopedia, magazine and newspaper articles. Perhaps the reader already knows that the 11th edition of the *Encyclopaedia Britannica* is not just an encyclopedia, and that *Sky and Telescope* is not just a magazine. By relying on these secondary sources I necessarily and unwittingly pass on some misconceptions, but the reader can at least check my sources to see if I have done my homework. I invite readers to help me set the facts straight, should there be a subsequent edition of this book. Otherwise we may have a situation like that in the movie *The Man Who Shot Liberty Valance* where a newspaper editor lectures Jimmy Stewart about truth: 'When the legend becomes fact, print the legend'.[3] Though professional historians would prefer that I cite and quote the originals, for this book it would require you and I to have a reading knowledge of Greek, Latin, Arabic, Turkish, French, German, Russian and English, and to have access to sources as obscure as the *Communications of the Yekaterinoslav Mining Academy*.

Concerning contemporary astronomy, I have shown how astronomy has become such a big field that it would be difficult to produce a readable and interesting book that covered everything. Writing about modern observatories is like planning a wedding: you want to keep it small and simple, but the list of invitees keeps getting longer and longer as a result of the refrain, 'So and so will be insulted if you don't invite them'. As anyone who has been to a big reception can testify, if it is too large, it is impossible to get to know anybody at all well. I hope that the reader finds my young and old 'guests' interesting and that sufficient paths have been provided for getting to know them better.

Acknowledgements

I gratefully acknowledge the significant contributions of Alan H. Batten (Dominion Astrophysical Observatory), Donald E. Osterbrock (Lick Observatory), Thomas R. Geballe (United Kingdom Infrared Telescope) and Frank K. Edmondson (Indiana University) who read portions of the manuscript and made numerous constructive comments. Owen Gingerich (Harvard) was his usual source of information and inspiration. Mitsuo Akiyama and Alika Herring were most helpful in reconstructing the events of the establishment of Mauna Kea as an astronomical site. The following librarians were particularly helpful: Ethleen Lastovica (South African Astronomical Observatory), Judy Lola Bausch (Yerkes), Dorothy Schaumberg (Mary Lea Shane Archives of the Lick Observatory), Gaila Vidunas (University of Hawaii, Hilo), and Sara Thompson (Hawaii County Library). I must also thank Douglas K. Duncan (Mount Wilson and Las Campanas Observatories), Riccardo Giacconi (Space Telescope Science Institute), Carlton M. Gillespie Jr (NASA/Ames Research Center), Ewen Whitaker (Lunar and Planetary Lab) and Dieter B. Herrmann (Archenhold-Sternwarte, Berlin).

I thank many individuals and institutions that provided figures for reproduction (who are acknowledged throughout the book), but I must single out Michael Broyles for being particularly helpful. Finally, I thank Dana K. Bairey for her encouragement and support.

A note on nomenclature

For the most part I have used the spelling of proper names adopted in the sources referenced in the footnotes. For the first two chapters I generally adopt the spelling given by George Sarton in his *Introduction to the History of Science*. The exceptions to this rule are 'Hipparchus' and 'Aristarchus', whose names are generally transliterated in this incorrect fashion.

Each chapter is a self-contained unit as far as the footnotes go. Therefore, a source such as Helen Wright's biography of George Ellery Hale appears as Ref. 1 of Chapter 6 and as Ref. 2 of Chapter 7.

I have translated some passages from the original German or Russian. The original Pulkovo statutes are given in Wilhelm Struve's 1845 *Description* in both French and Russian. I thank Alan Batten for clarifying some passages (originally in French) from that work.

1

The Alexandrian Museum

There it towered up, the wonder of the world, its white roof bright
against the rainless blue; and beyond it, among the ridges and
pediments of noble buildings, a broad glimpse of the bright blue sea.
Description of the Alexandrian Museum.
(Charles Kingsley, *Hypatia*)

Alexander the Great (356–323 BC), student of Aristotle and wielder of
empires, founded the Egyptian city of Alexandria in 332 BC. Two years
previously he had begun his campaigns against the Persian Empire. Egypt
was the last of the Mediterranean provinces to be won, and Alexander
decided to build his capital there, at the mouth of the westernmost branch of
the Nile. Alexander stayed in Egypt less than a year. He honored the Egyptian
religion by making a pilgrimage to the oracle of Zeus Ammon, where he was
declared son of the god. He placed the taxation and control of the Egyptian
army and navy in the hands of trusted Greeks. In 331 BC he sailed away,
leaving the further establishment of the new city in the hands of one
Cleomenes. Alexander died in Babylon at the age of 33 but was eventually
entombed in Alexandria.

After Alexander's death, Egypt was ruled by the Ptolemies, Greeks whose
line died out with Cleopatra in 30 BC. (They were not related to Claudius
Ptolemy, whom we shall meet later in this chapter.) The first king of this line,
known as Ptolemy Soter, had been one of Alexander's generals. He promptly
did away with Cleomenes and spent the better part of a 40-year reign
consolidating his power in the Mediterranean. Under his son, Ptolemy II,
surnamed Philadelphus, Alexandria became a great city, and the court
became comparable to the Versailles of Louis XIV.

Originally, the city of Alexandria consisted of nothing more than the
island of Pharos, connected to the mainland by a stone wall known as the
Heptastadium, and the fishing village of Rhakotis. It was on the island that the
first two Ptolemies constructed the famous lighthouse, one of the Seven
Wonders of the World. The mainland city soon contained three principal
regions, the Jewish quarter (northeast), the Egyptian quarter (west; Rhakotis), and the central Greek section (Brucheum). There is ample evidence that
the metropolitan mixture of people who lived in Alexandria and those who

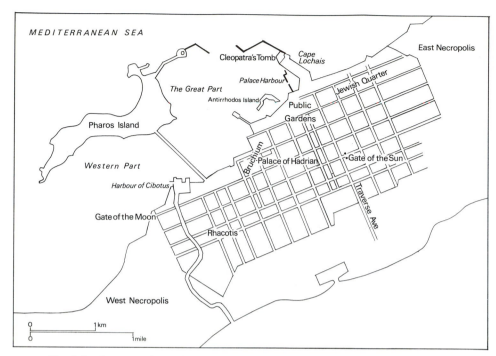

Fig. 1.1. Ancient Alexandria. After Willis, F. Roy, *World Civilizations, vol. I, From Ancient Times Through the Sixteenth Century*. Lexington, Mass.: D.C. Heath, 1982, p. 161.

visited there got on well with each other. The Ptolemies followed Alexander's example and sought to combine aspects of Greek and Egyptian religion. The population of Alexandria in 30 BC is estimated to have been 300 000 free men and women, and a large number of slaves. To be more exact, everybody got along reasonably well until the third century AD, which saw the persecution of the general populace by the Romans, then, beginning in the fourth century, the persecution of the Jews and other non-Christians by the Christians.[1]

Begun by order of Ptolemy Soter under the direction of Demetrios of Phaleron and constructed during the time of the first two Ptolemies, the Alexandrian Museum (or temple of the Muses) derived from Aristotle's Lyceum near Athens. It was a research institution dedicated to the pursuit of excellence in literature, medicine, mathematics, and astronomy. Many famous scholars were associated with the Museum, among them: Zenodotos of Ephesos, the first librarian of the great library; Callimachus, who helped organize the library at the Serapeum (temple of Serapis); Eratosthenes (librarian from 235–195 BC), who determined that the Earth was spherical and had a diameter of 7850 miles; Aristophanes of Byzantium (librarian from 195 to 180 BC), considered the inventor of punctuation; and Aristarchus of Samothrace, who succeeded Aristophanes as librarian of the Museum and was a noted philologist (not to be confused with Aristarchus of Samos, discussed below). With the decline of Greece, Alexandria soon became the center of learning in the Western World. In a short time the collection of works at the Museum (the 'mother' library) grew so large that a separate ('daughter') library at the Serapeum was built. The number of scrolls accessible to scholars in Alexandria amounted to several hundred thousand (Fig. 1.2).[2]

The Muse of astronomy, Urania, was alive and well in Alexandria. One of the most famous ancient astronomers was Aristarchus of Samos, who observed in Alexandria from 280 to 264 BC. He was the first to propose the heliocentric hypothesis – that the Earth revolved about the Sun. He has been dubbed the 'ancient Copernicus', pre-dating the famous Pole by 1800 years. He also proposed a method of determining the distances to the Moon and Sun which was geometrically correct but beyond the capabilities of the measuring instruments of the day. His value for the distance to the Sun, though slightly modified from time to time, was generally accepted until the seventeenth century, when it was shown to be twenty times too small.*

Our primary source of information on Greek astronomy is the famous work by Claudius Ptolemy known as the *Almagest*. The original Greek title was ἡ μαθηματικὴ σύνταξις [*E mathematike Syntaxis*: mathematical (in this case meaning astronomical) compilation or treatise]. It later became known as ἡ μεγάλη σύνταξις or ἡ μεγιστη σύνταξις [the great/*megale* (or greatest/*megiste*) compilation/*Syntaxis*]. One of the first Arabic translators of the work, al-Ḥajjāj ibn Yūsuf (*c.* AD 829) derived an artificial contraction of the words μεγάλη σύνταξις to obtain his title *Kitāb* (book) *al-mijisti*, hence *Almagest*.[5] The significance of this book in the history of astronomy cannot be over-estimated. As has been said, 'The importance of a scientific work can

---

* The ancient value of the solar parallax of 3′ implied that the Sun's distance is about 1200 Earth radii. The actual value of the parallax is 8″.8 (corresponding to a distance of 92.9 million miles, or about 23 400 Earth radii). It is curious that the only ancient parameter which Tycho Brahe accepted was that of the solar parallax, and that turned out to be wrong (see Chapter 3). The first evidence that the value was on the order of 10″ was obtained by J. D. Cassini I in Paris and by John Flamsteed from observations at Derby, both from observations of Mars in the 1670s.[3,4] (A circle contains 360°. 1° = 60′ (arc minutes). 1′ = 60″ (arc seconds). 1° therefore equals 3600″.)

Fig. 1.2. Model of the Alexandrian Museum. (Photographs courtesy of Carl Sagan.)

be measured by the number of previous publications it makes superfluous to read'.[6] Though the survival of the *Almagest* and the oblivion of other works not handed down may have been a matter of fate, it is more likely that after its compilation the *Almagest* was recognized as the truly comprehensive summary of Greek astronomy that it is, and sufficient care was exercised to ensure its survival. Though opinions vary as to the originality of Claudius Ptolemy's personal contributions,[7-10] it is an historical fact that the *Almagest* was the Bible of astronomy from the time of its compilation (about 150 AD) until Copernicus published *De revolutionibus* (1543), and we are fortunate that an institution such as the Alexandrian Museum existed to help preserve the evidence of the state of Greek astronomy.

We can provide here only a general review of Greek observational accomplishments and the Ptolemaic theory of the heavens.[11] The two most significant astronomers of ancient times were Hipparchus of Nicaea (d. *c.* 127 BC) and Claudius Ptolemy (*c.* AD 100–170). Separated by almost three centuries, what we know of the first is almost wholly derived from the second, and the fame of the second as an astronomical model builder is very largely based on the accomplishments of the first in the realm of observational astronomy. Hipparchus observed from the island of Rhodes from 147 (perhaps 162) to 127 BC. He was not directly associated with the Alexandrian school.[12] Among his accomplishments, based on his own observations and those of other Greeks as well as Babylonians, are the determinations of:

1) The length of the tropical year (the amount of time from one spring equinox to another). He found it to be $365\frac{1}{4} - 1/300 = 365.2467$ days. The modern value is $365.2422 \approx 365\frac{1}{4} - 1/128$ mean solar days.

2) The length of the sidereal year (the amount of time it takes the Earth to go through 360° of its orbit with respect to the stars, though for 'Earth' we should have written 'Sun' as the Greeks considered the Earth to be fixed). He found $365\frac{1}{4} + 1/144 = 365.2569$ days. The modern value is 365.2564 days.

3) The lengths of the (northern hemisphere) seasons: spring $- 94\frac{1}{2}$ days; summer $- 92\frac{1}{2}$ days; autumn $- 88\frac{1}{8}$ days; and winter $- 90\frac{1}{8}$ days.

4) The synodic periods of

Mercury: 46/145 years = 115.87 days (modern value 115.88 days),
Venus: 8/5   years = 584.39 days (modern value 583.92 days),
Mars: 79/37 years = 779.85 days (modern value 779.94 days),
Jupiter: 71/65 years = 398.96 days (modern value 398.88 days),
Saturn: 59/57 years = 378.06 days (modern value 378.09 days).

5) The obliquity of the ecliptic (23° 51′; actual value for 150 BC, based on modern theory, was 23° 42′; modern value is 23° 27′).

6) The inclination of the Moon's orbit to the ecliptic (5°; modern value 5° 09′).

7) The Moon's horizontal parallax, corresponding to a mean distance from the Earth inbetween 59 and $67\frac{1}{3}$ Earth radii (modern value 60.26).

8) The place of the Sun's apogee (greatest distance from Earth), celestial longitude $65\frac{1}{2}°$. This changes $61''.8$ per year and in 1987 is $102°\,42'$.[13]

9) The eccentricity of the 'Sun's orbit' ($\frac{1}{24} = 0.042$; modern value $0.016722 \approx 1/60$).

10) The eccentricity of the Moon's orbit, which is inferred from the 'first inequality of the Moon's motion in longitude'.

11) The precession of the equinoxes ($36''$/year; actual value in 150 BC was $49''.8$/year; modern value is $50''.3$/year).

12) He also compiled a catalogue of 850 stars.

Lest the reader be intimidated by these technicalities, let us offer some explanation and definitions. In spite of the suggestion of Aristarchus of Samos that the Sun was the center of the solar system, Greek astronomy was founded on the notion that all bodies revolved (orbited) about the Earth. The apparent orbit of the Sun about the Earth traces out the *ecliptic*, which in modern terms is the intersection of the plane of the Earth's orbit and the celestial sphere. The angle between this plane and the Earth's rotational axis is called the *obliquity of the ecliptic*. Hipparchus knew that the Sun was not always at the same distance from the Earth, hence the eccentricity of the 'Sun's orbit about the Earth'. However, orbits were considered to be composed of circular components, so the term eccentricity relates to how far the Earth would be displaced from the center of the 'Sun's orbit'.

The planets (i.e. 'wanderers') were star-like objects seen to move against the background of stars. For the most part they moved from west to east against the background of stars. This is known as *direct motion*. The Sun, Moon, planets and stars appear to rise and set each day (owing to the Earth's rotation). The Sun appears to move smoothly eastward along the ecliptic (though at a slightly variable speed), but the planets can undergo *retrograde* (backwards, i.e. from east to west) motion against the stars. The amount of time from the occurrence of retrograde motion of a planet to the next such occurrence for that planet is known as its *synodic period*. If you can imagine all the planets orbiting the Sun, and place yourself on the Earth, you can imagine that retrograde motion occurs when the Earth, a planet, and the Sun are all in a line, and the planet is at its closest approach to the Earth. If you were riding along above the Earth's north pole with Mars ahead of you, due to Mars' motion (counterclockwise around the Sun) it will appear to be moving from right to left (which would be west to east) against the stars. Mars appears more and more on your right side as the Earth, moving faster in the same counterclockwise direction, starts to overtake it. As the Sun (on

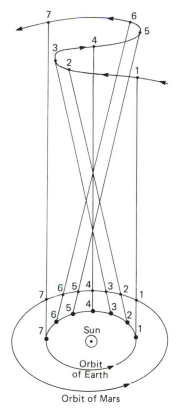

Fig. 1.3. Retrograde motion is the apparent backwards movement of a planet against the background of stars due to the unequal rates of revolution of the Earth and planet about the Sun. From the *Cambridge Encyclopaedia of Astronomy*, p. 163, Fig. 9.5.

your left), the Earth, and Mars (now directly on your right) fall into a line, you will observe Mars to move at its maximum apparent speed from left to right against the distant stars. All this is understandable with the modern heliocentric (Sun-centered) arrangement of the solar system. However, the Greeks were only concerned with the apparent positions of the planets in the sky; they derived a model to explain the appearances, given the assumption of a geocentric (Earth-centered) system.

More easily understandable than synodic period is the *sidereal period* (i.e. period by the stars). Once again, if we may invoke the notion of a bird's eye view of the solar system, the sidereal period is the amount of time it takes a planet (including the Earth) to orbit the Sun or the amount of time it takes a satellite to orbit a planet, with the position of the planet (or satellite) in its orbit referred to the distant stars. In other words, it is the amount of time it takes a body to traverse $360°$ of its orbit about its parent body, using a distant ('absolute') frame of reference.

Another important concept is that of *parallax*. This is the difference in direction towards a body depending on the place of observation. If you observe the direction towards a tree and move 10 feet to the right and observe that the tree is in a different direction (using, say, a magnetic compass), the difference in direction is the *parallax angle*. Given that angle and the size of the baseline (10 feet), with simple geometry one may calculate the distance to the tree. This simple concept can be used to calibrate the universe in relative terms (the number of baselines), or, if the baseline can be expressed in some other known units (feet, or miles), the universe can be calibrated in absolute terms. For example, using observations of planets, Copernicus was able to determine the relative sizes of the planetary orbits, using the Earth–Sun distance (1 astronomical unit, or AU) as his standard. But he did not know how many miles were in an astronomical unit. From observations of the position of the planet Mars with respect to nearby stars in the sky over the course of a night, Flamsteed and Cassini in the late seventeenth century were able to determine the distance to Mars in absolute terms using a portion of the Earth's diameter as a baseline, and hence, to determine the sizes of all the planetary orbits as well. When we speak of the *solar* (or lunar) *parallax* we mean the apparent difference in direction towards the Sun (or Moon) given its distance from us and given the radius of the Earth. For the Moon this is just under $1°$ (averaging $57'$): for the Sun it is $8''.8$. The former was easily measurable given the instrumentation available to the Greeks, whereas the latter was not, which explains why the ancient Greeks' value of the distance to the Moon is in accord with the modern value and why an accurate determination of the Sun's distance had to wait until the invention of such things as telescopes and micrometers.

The final use of the term parallax pertains to the distances to the stars. Assuming that the Earth revolves about the Sun, the nearby stars should exhibit shifts in their positions with respect to the distant background stars. The shift would have a period of exactly one year. The baseline is, of course, the radius of the Earth's orbit. Since the distances to the stars are immense, the corresponding *trigonometric stellar parallaxes* are very small angles (less than $1''$). We shall discuss this further in Chapters 3, 4, and 5.

Regarding celestial coordinates, there are three principal systems to consider. As the sky appears to be an inverted bowl above us, and we are on the inside of the *celestial sphere*, any point on the surface of this sphere can be described by specifying two coordinates. In the *horizon* system we use the *altitude* (or elevation angle – the number of degrees an object is above the

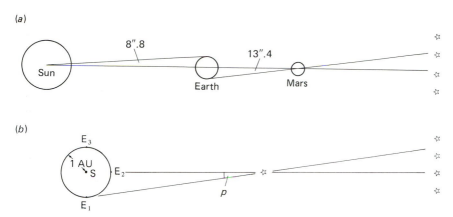

Fig. 1.4. Principles of parallax. The 'solar parallax' is indicative of the distance to the Sun, using the radius or diameter of the Earth as a baseline. The 'planetary parallax', using the Earth as a baseline, can be used to determine the distance to other objects in the solar system because the relative sizes of the orbits are known from Kepler's third law. 'Stellar parallax' is indicative of the distances to the nearby stars, using the radius or diameter of the Earth's orbit as a baseline and the distant background stars or background galaxies as the fixed frame of reference. (*a*) Using the Earth as a baseline to determine solar parallax or parallax of Mars. (*b*) Using the Earth's orbit as a baseline to determine the parallax of a nearby star. S, Sun; E_1, E_2, E_3, positions of the Earth. If $p = 1''$, distance to the star is $206\,265$ AU $= 1$ parsec $= 3.26$ light-years.

horizon) and the *azimuth*. If we extend an arc from the *zenith* (straight up) down through the object, continuing perpendicular to the horizon, then the number of degrees from north, measuring clockwise around the horizon to this point, is the azimuth angle. We also speak of the zenith angle, which is $90°$ minus the elevation angle. The *celestial meridian* is the great circle passing from the south point on the horizon to the zenith, through the celestial pole and down to the north point on the horizon.

As the Earth is spinning on its axis (or at least it looks like all the celestial objects are moving across the sky), coordinates in the horizon system change continuously. Astronomers prefer coordinates in some 'fixed' frame of reference. For *ecliptic coordinates* the number of degrees an object is north or south of the ecliptic is the *celestial latitude* (β). Making an arc parallel to the ecliptic, the number of degrees east of a defined reference point is the *celestial longitude* (λ). The reference point is the *vernal equinox* (Υ), which is the position of the Sun on the ecliptic about March 21.

The third system is that of *equatorial coordinates*. The axis of rotation of the Earth intersects the celestial sphere at the north and south celestial poles. The plane of the Earth's equator extended to the celestial sphere defines the

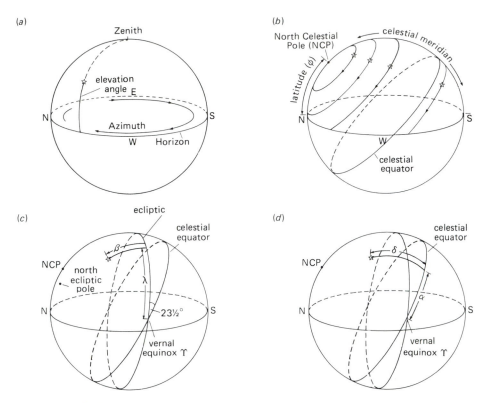

Fig. 1.5. The principal systems of celestial coordinates (for an observer in the northern hemisphere). (*a*) Horizon system. (*b*) Daily paths of stars for latitude ϕ. (*c*) Ecliptic coordinates; celestial longitude (λ) and celestial latitude (β); (*d*) Equatorial coordinates; right ascension (α) and declination (δ).

celestial equator. The number of degrees an object is north or south of the celestial equator is the *declination* (δ). The analog of longitude on the Earth is *right ascension* (α) in the sky, which is the number of degrees (or hours, minutes and seconds, since $360°$ corresponds to $24\,\mathrm{h}$) an object is east of the zero point of right ascension, which, as in the ecliptic system, is the vernal equinox. The vernal equinox is then the point of intersection of the ecliptic and the celestial equator where the declination of the Sun changes from negative to positive.

One's geographical latitude (ϕ) is simply the elevation angle of the north celestial pole. The declination corresponding to the observer's zenith is also equal to his latitude. If the observer is located in-between the Tropic of Cancer ($\phi = 23\frac{1}{2}°$ N) and the Tropic of Capricorn ($\phi = 23\frac{1}{2}°$ S), twice a year the Sun will pass through the zenith at noon. These tropical boundaries are defined precisely by the value of the obliquity of the ecliptic.

Returning to ancient Greece, Hipparchus' greatest discovery was that of the precession of the equinoxes. This is the westward movement of the vernal equinox along the celestial equator. From records of star positions by other Greek astronomers (notably Timocharis of Alexandria, *c.* 300 BC) and Babylonians, Hipparchus noted that the celestial longitudes of the stars tended to increase with time, about 1° per century (36″/year), while the celestial latitudes remained the same. We know now that this effect is due to the torque of the Moon and Sun on the spinning, oblate Earth. The modern value of precession (50″.3/year) indicates that the vernal equinox slides around the ecliptic with a period of 25 800 years. That means that the Pole Star is not always the same star and that the right ascensions and declinations of all the stars change. Thus stellar coordinates are said to have an epoch (e.g. 'epoch 1950'). Also, because of precession, the vernal equinox – known as the First Point of Aries – no longer resides in Aries, but in Pisces. The positions of the summer and winter solstices are similarly rotated around the zodiac. To be more exact, the Tropic of Cancer could now be renamed the Tropic of Gemini, and the Tropic of Capricorn could be called the Tropic of Sagittarius.

Originally, the Greeks had developed models of the universe based on the notion of 'homocentric' (i.e. geocentric) spheres. Eudoxus of Cnidus (*c.* 400–347 BC) derived a system of 27 spheres. He imagined that each planet was carried along on a large sphere. To account for non-uniform motion and variations of motion, more than one sphere per body had to be used. This required four spheres each for Saturn, Jupiter, Mars, Mercury, and Venus, three each for the Sun and Moon, and one for the stars. As these 'crystalline' spheres revolved about the Earth, they generated the 'music of the spheres'. Callippos of Cyzicos (born *c.* 370 BC) increased the number of spheres to 34, while Aristotle (384–322 BC) used 55. Apollonios of Perga (born *c.* 262 BC), who flourished in Alexandria in the second half of the third century BC, introduced notions which gradually replaced the idea of homocentric spheres. To account for the principal (direct) motion of a planet he invented the concept of a *deferent*, which was a perfectly circular orbit about the Earth. As viewed from above the Earth's north pole, all the deferents turned counterclockwise. To account for retrograde motion Apollonios described the *epicycle*, which was a smaller circle whose center rode along on the deferent. The epicycle turned faster than the deferent such that when the planet (moving along on the epicycle) was moving in the direction opposite to that of the deferent, the planet is seen to move from east to west against the stars. Either Apollonios or Hipparchus devised the notion of *eccentrics*. If the Earth were at the center of a deferent, it was said to be a concentric deferent;

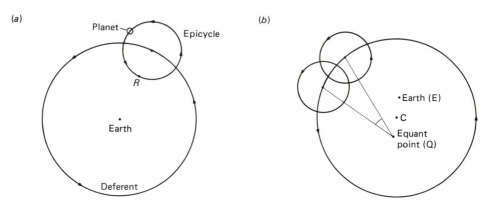

Fig. 1.6. Representations used in the geocentric solar system models of the Greeks. (*a*) The center of the epicycle moves uniformly along the deferent; when the planet is at position R it is observed to undergo retrograde motion. (b) Here the Earth (E) is placed eccentrically with respect to the center of the deferent (C). The deferent turns uniformly with respect to the equant point. Distance EC = CQ.

if the Earth were displaced from the center, the deferent was said to be eccentric. This would account for the variable speed (and apparent brightness) of the Moon, Sun, or a planet as viewed from the Earth.

Ptolemy's *Almagest* provides the most advanced Greek model of the universe. He retained the notions of deferent, epicycle, and eccentric, and also introduced the notion of the *equant* point, which is a variation of the concept of the eccentric. Here the Earth is displaced from the center of the deferent, but we imagine a point on the opposite side of the center from the Earth and at the same distance from the center as the Earth; the motion of the center of the epicycle on the deferent is uniform (in degrees per day) as seen from the equant. By invoking a number of epicycles and by determining their speeds, the speed of the deferent with respect to its center or with respect to an equant point, and the eccentricity of the position of the Earth, it was possible to produce a mathematical model to match the observed positions of a planet (or the Moon or Sun) and to predict the future positions. However, the reader will see the major philosophical flaw in all this: if discrepancies are found between theory and observation, all one has to do is adjust the parameters or add more epicycles. Geometrical models were produced for the motion of the Sun, Moon, and planets, but they quickly became more complicated than the sets of homocentric spheres. Also, they were mathematical models, not necessarily (or likely to be) physical representations of the actual motions.

The star catalogue of Hipparchus had been a mixed bag of information. It was, at minimum, a list of the numbers of stars in the constellations and some of the relative positions of the stars. It is probably true that Hipparchus measured coordinates for a sizeable number of stars, though these were a combination of ecliptic and equatorial coordinates. For a north–south coordinate he first used declination (or the polar distance $90° - \delta$). For an east–west coordinate he sometimes gave right ascensions but more often gave a kind of 'polar longitude' (the celestial longitude corresponding to the intersection of the declination circle and the ecliptic).[14] After the discovery of precession Hipparchus introduced the use of ecliptic coordinates because, while the celestial longitudes would change with time, the celestial latitudes would remain the same.

In books VII and VIII of the *Almagest*'s 13 books, Ptolemy discusses the fixed stars. He first re-derives a value for the precession. Using observations of 18 stars at the times of Timocharis, Hipparchus and in his own day, he seems to have found that 6 of them gave a value of precession around 38″.1/year, while the other 12 would have given 52″.8/year. The weighted mean would be 47″.9/year, close to the actual value at that time of 49″.8/year. However, because the value of 38″.1 for 6 stars closely matches Hipparchus' value of 36″, Ptolemy apparently decided that the values of the precession derived from the other 12 stars were wrong. He then adopted Hipparchus' value of 36″/year, a decision that was carried out with a certain degree of consideration rather than by blind copying.[15]

I have mentioned these details regarding the star catalogue of Hipparchus and 'Ptolemy's value' of precession because of the ongoing controversy concerning whether or not Ptolemy was a great compiler, theoretician, and observer in his own right, or a fraud who simply made up data to fit his theories. This debate arose with Tycho Brahe (*c.* 1598),[16] was rehashed by Delambre (*c.* 1817),[17] and most recently pursued by R. R. Newton (1977).[18]

Ptolemy's star catalogue contains 1022 stars, with their respective celestial latitudes and longitudes (at epoch AD 138), and also their apparent magnitudes. His longitudes are systematically 1° too small, *as if* he had taken a previous catalogue with coordinates at the epoch of Hipparchus 265 years before and precessed them to his own day using an incorrect value of precession, adding $2\frac{2}{3}°$ instead of $3\frac{2}{3}°$.* However, advocating the notion that Ptolemy simply borrowed the star catalogue of Hipparchus ignores a number of points, among them: (*a*) Hipparchus probably did not give coordinates for all his stars. Though Ptolemy could also have obtained some observations

---

* Here 265 years × 36″/year = 9540″ = 2° 39′ and 265 years × 50″/year = 3° 41′.

from Menelaos (c. AD 98), it is likely that Ptolemy made a sizeable number of his own observations. (b) Hipparchus did not uniformly use celestial coordinates. For Ptolemy to 'borrow' the celestial latitudes he would have had to convert many of Hipparchus' declinations. To 'simply precess' Hipparchus' longitudes he would at least have had to convert a large number of polar longitudes to celestial longitudes. (c) Hipparchus' list contained observations of no more than 850 stars,[19] implying that Ptolemy had to have done some of his own observing for 172 stars. He had to do more than just steal a catalogue of Hipparchus.[20]

There are many outstanding questions regarding Ptolemy's star catalogue. For example, Alexandria is 5° further south than Rhodes (where Hipparchus observed), but Ptolemy's catalogue does not contain any stars that are visible at Alexandria while being invisible further north. However, because the catalogue of Hipparchus has been lost, we may never know for certain to what extent Ptolemy borrowed from Hipparchus. Still it can not be denied that Ptolemy's star catalogue, indeed the entire *Almagest*, was the result of a great effort to synthesize and compile Greek astronomical knowledge.

A final piece of information about Greek observational theory which we will mention deals with the motion of the Moon. The Babylonians had left particularly good records of lunar eclipses dating back to 747 BC. These could be used for predicting future eclipses because the periodicity of the Moon's orbit had been well established. For example, the mean synodic period of the Moon (the amount of time from one full moon to the next or from one new moon to the next) was determined to be the equivalent of $29^d \, 12^h \, 44^m 3^s.3$ (within 1^s of the value for that time based on modern theory). Hipparchus discovered that the Moon moved most rapidly at one point in its orbit and most slowly at the opposite point. He envisioned the Moon moving along on a deferent with the Earth placed eccentrically with respect to the center of the deferent. He also found that the line joining the Earth and the center of the Moon's orbit (the 'line of apsides') was not fixed in space. It pointed toward succeeding signs of the zodiac and completed a revolution in just under nine years. But the theory of the Moon's motion at that time only predicted the correct position of the Moon when it was opposed to the Sun (full moon) or in conjunction with the Sun (new moon). Ptolemy discovered that when the Moon was at quadrature (first or third quarter moon) the observed longitude was different than that predicted by the simple eccentric model. The error could amount to $1°.27$. This is known as the *evection*. In order to explain this, Ptolemy invented a 'crank mechanism' whereby the center of the Moon's deferent was cranked around the Earth once a month (in addition to the much slower rotation of the line of apsides). A consequence of Ptolemy's lunar theory is that it predicted that the greatest Earth–Moon distance was 70% larger than the minimum Earth–Moon distance. The measuring instruments of that day, such as the staff of Hipparchus (see below) showed that the Moon's distance did not vary by this large amount.[21]

Though Ptolemy's theory of the Moon's motion was an improvement on what had been done by the Babylonians and Hipparchus for the prediction of the longitude of the Moon, it was a far cry from being complete. Tycho Brahe discovered four more inequalities in the Moon's motion (see Chapter 3). The development of the theory of the Moon's motion played a large role in the history of Greenwich Observatory (Chapter 4). Due to the complexities of the Moon's orbit, improvements to the theory of the Moon's motion are being made even today.

The instrumentation used by the Greeks ranged from the very simple to the somewhat complex.[22] Their instruments may be divided into four basic types: (*a*) those for reckoning time (determining the passing of the seasons or the time of day); (*b*) instruments for determining the angular size of the Sun or Moon; (*c*) instruments for determining the elevation angle of any object above the horizon; and (*d*) more complicated instruments for determining the celestial coordinates of any object (as opposed to coordinates in the horizon system).

In the first group we must first of all mention the *gnomon*, which was a vertical stick placed in the ground. With it one could determine the north–south line by making a mark on the ground at the end of the shadow of the stick at some time in the morning, then making a mark in the afternoon when the stick's shadow was of equal length. The north–south line is then drawn from half way between these two points to the gnomon. Then, on any subsequent day the moment of apparent noon is simply the time that the shadow of the stick falls on the north–south line. From the length of the shadow one can determine the approximate first day of summer or the first day of winter by noting when the shadow at noon was the shortest or longest, respectively. The idea of the gnomon led to the development of sundials.

A much later device for determining the seasons was the *solstitial armillary*, which consisted of a graduated metal ring placed vertically in the plane of the meridian, within which a smaller ring turned. The inner ring had two knobs on opposite sides of a diameter. The shadow of one was made to fall on the other by turning the ring. A pointer attached to the lower knob allowed one to read off the Sun's elevation angle on the graduated outer ring. As with the gnomon, when the Sun was highest in the sky at noon it was the first day of summer.*

An *equinoctical armillary* or *equinoctial ring* was used for determining the equinoxes, as opposed to the solstices. It consisted of a flat ring mounted parallel to the plane of the Earth's equator. A perpendicular line through the center of the ring would point toward the celestial pole. When the shadow of the front (upper) half of the ring exactly covered the inside edge of the back (lower) half of the ring, it was the first moment of spring or the first moment of autumn.

Instruments of the second group include the *staff of Archimedes* and the *staff of Hipparchus.* The former is a rigid stick upon which an occulting disk can slide. By aiming it at the Sun or Moon, from the size of the disk and its distance from the eye along the stick one can determine the angular diameter of either object. The staff of Hipparchus is almost the same, except instead of an occulting disk it uses a piece of wood with a sizeable hole drilled through. With these devices the Greeks were able to show that the distance between the Earth and Moon changed by a small amount.

Ptolemy realized that the whole circle used for the solstitial armillary was not needed to obtain the elevation angle of an object. If one used a quarter of a circle one could build an instrument of the same physical size which could be calibrated to smaller angles. This allowed more accurate data to be obtained. Thus was born the *quadrant*, which was made of wood, stone, or metal. Quadrants were used (with various improvements) through the eighteenth century. They were principally used for the measurement of elevation angles but the smaller ones were sometimes mounted so as to be rotatable about a vertical axis, in which case they could be used to measure azimuths as well.

A second instrument in the third group is the *triquetrum* (also known as Ptolemy's rulers or a parallactical instrument), which consisted of a solid vertical post with one peg at the top and another peg closer to the ground. A second stick equal in length to the distance between the two pegs pivots from the top, while a third stick pivots from the bottom peg. The bottom of the second stick can slide along the length of the third stick. Sights on the second stick allow one to aim it at some object (usually the Moon) in the sky, and the position of the bottom end of the second stick fixes a place on the third stick.

---

* To be more exact, it was the moment of the summer solstice in the northern hemisphere when the declination of the Sun reached its maximum positive value. If one is situated in-between the equator and the Tropic of Cancer, the Sun will pass north of the zenith.

The three sides form an isosceles triangle, the base of which is the distance from the vertical post along the third stick to the lower end of the sight-carrying stick. From the shape of the triangle (calibrated by means of graduations on the third stick) one can determine the zenith angle of the object. Copernicus owned one of these and Tycho Brahe built one as well (see Fig. 3.2).

The fourth group of instruments contains *armillary spheres* and *astrolabes* (i.e. star finders), either spherical or plane. An armillary sphere is a set of rings (*armillae*) used to represent such celestial great circles as the meridian, horizon, ecliptic and celestial equator. With the meridian ring placed appropriately along the north–south line and rotated in that plane to correspond to the observer's latitude, it was possible to use sights on the armillary sphere directly to read off the celestial latitude and longitude or right ascension and declination of an object. An armillary sphere of four rings was probably used by Hipparchus, but after him no further progress was made in their construction until the time of Tycho Brahe (sixteenth century). The discussion of astrolabes shall be left until the next chapter, as the Arabs were more avid users and makers of them (in particular the plane astrolabe).

Just as Greek astronomy reached its height with Ptolemy's compilation of the *Almagest*, after which we must look carefully for astronomers of note, so the splendor of Alexandria reached the beginning of the end of its halcyon days. As part of the Roman Empire and one of the centers of Christian ecclesiastical power, Alexandria soon became the scene or origin of political and religious upheavals. In AD 215, the Roman emperor Caracalla ordered the massacre of all Alexandrian youths capable of bearing arms, to avenge some unflattering comments aimed at him by the local inhabitants. This being a survivable perturbation of external origin, the city eventually recovered its bearings, but internal religious controversy began to grow. About 318 Arius (*c.* 280–336) began to preach the idea that Christ was of a *similar essence* (ὁμοιούσιοζ) to God, while Athanasius (293–373), also of Alexandria and later bishop there, advocated the notion that Christ was *consubstantial* (ὁμοούσιοξ) with God. This may seem to be no basis for a major schism, *as the difference rests on a single iota*, but it precipitated the Council of Nicaea, ordered by the Byzantine emperor Constantine in 325, at which Catholic beliefs were carefully defined. Arianism was declared to be heresy, but after the death of Constantine in 337, the new emperor reversed this decision, making Arianism the orthodox doctrine until 378. Many of the barbarian tribes, such as the Goths and Vandals, were converted to the faith as Arians.[23]

The Byzantine emperor Theodosius, who ruled from 379 to 395, published various edicts against paganism, which resulted in the persecution of those of unorthodox belief (i.e. the Jews and those who favored the doctrine of Arius). These edicts also led to the destruction of many works of scholarship and the general decline of scholarship. In 389 or 390 Theodosius issued an order for the destruction of the great statue of Serapis at Alexandria, which had been created during the reign of Ptolemy Soter and which embodied the attempts on the part of the Ptolemies to combine elements of Greek and Egyptian religion. In the fourth century AD this was deemed no longer necessary. The damage done to the Serapeum may have also resulted in the destruction of its library, by then numbering a few hundred thousand volumes. Many great works of Greek scholarship may have been irrevocably lost in that wave of the destruction of paganism, although it is the opinion of a number of writers that the 'daughter' library at the Serapeum survived until the Saracen conquest in the seventh century.[24,25]

Scholarship did continue in Alexandria. Whatever the fate of the library at the Serapeum, the 'mother' library of the Museum was still intact,[26] and where there are books and scholars, there will be scholarship. The most important mathematician and astronomer at that time was Theon of Alexandria (fl. *c.* 365–400). He composed a commentary on the *Almagest*. Theon's revised edition of the *Handy Tables*, which were tables for astronomical computation extracted by Ptolemy from the *Almagest*, has survived. (Ptolemy's original version has not.) These tables were used by the Arabs and were later embodied in the Toledan Tables of the eleventh century (see p. 41). During the Middle Ages, Theon was thought to be the main author of Euclid's geometry. His treatise on the plane astrolabe was the principal means of transmission of knowledge of that instrument to the later Arabic scholars.

Better known, but only because of her gruesome death, was Theon's daughter Hypatia. Of her we know very little. According to Suidas (fl. *c.* middle of tenth century), she was the author of commentaries on the *Arithmetica* of Diophantus of Alexandria, on the *Conics* of Apollonios of Perga, and the 'Canon' (*Handy Tables*) of Ptolemy. These works have not survived. She may have helped her father Theon with his commentaries on the *Almagest*.

Socrates Scholasticus (*c.* 380–450), a contemporary of Hypatia, wrote an *Ecclesiastical History* covering the years 305 to 439, which provides us with the following:

> There was a woman at *Alexandria*, by name *Hypatia*: She was daughter
> to *Theon* the Philosopher: She had arrived to so eminent a degree of

learning, that she excelled (all) the Philosophers of her own times, and succeeded in that *Platonick School* derived from *Plotinus*, and expounded all the Precepts of Philosophy to those who would hear her. Wherefore, all persons who were studious about Philosophy, flockt to her from all parts: By reason of that eminent Gracefulness and Readiness of expression, wherewith she had accomplish'd her self by her Learning, she addressed frequently even to the Magistrates, with a singular modesty. Nor was she ashamed of appearing in a publick assemb[l]y of men. For all persons revered and admired her for her eximious modesty. Envy armed itself against this woman at that time. For, because she had frequent conferences with *Orestes* [the Roman prefect of Alexandria], for this reason a calumny was framed against her amongst the Christian populace, as if she hindered *Orestes* from coming to a reconciliation with the Bishop. Certain persons therefore of fierce and over-hot minds, who were headed by one *Peter* a Reader, conspired against the woman, and observe her returning home from some place. And having pulled her out of her Chariot, they drag her to the Church named *Caesareum*: Where they stript her, and murder'd her with [oyster] Shells [$\delta\sigma\tau\rho\alpha\kappa o\iota\zeta$ $\alpha\nu\epsilon\tilde{\iota}\lambda o\nu$]: And when they had torn her piece-meal, they carried all her Members to a place called *Cinaron*, and consumed them with fire. This fact brought no small disgrace upon *Cyrillus* [the bishop] and the *Alexandrian* Church. For, murthers, fights, and things of that nature, are wholly foreign to the embraces of *Christianity*. These things were done on the fourth year of *Cyrillus*'s Episcopate [AD415], in *Honorius*'s tenth, and *Theodosius*'s sixth Consulate, in the month of *March*, in *Lent*.[27]

To backtrack a moment, the bishop of Alexandria at the time of the sacking of the Serapeum was Theophilus, described by Gibbon as 'the perpetual enemy of peace and virtue; a bold, bad man, whose hands were alternately polluted with gold and with blood'.[28] In 412, Theophilus was succeeded by his nephew Cyril, who

soon made himself known by the violence of his zeal against Jews, pagans and heretics or supposed heretics alike. He had hardly entered upon his office when he closed all the churches of the Novatians [whom Gibbon describes as 'the most innocent and harmless of the sectaries'[29]] and seized their ecclesiastical effects. He assailed the Jewish synagogues with an armed force, drove the Jews in thousands from the city, and exposed their houses and property to pillage. The prefect of Egypt, Orestes, who endeavoured to withstand his furious zeal, was in turn denounced himself, and had difficulty in maintaining his ground against the fury of the Christian multitude ... [T]here can be no doubt that 'the perpetrators [of the murder of Hypatia] were officers of the church', and undoubtedly drew encouragement from [Cyril's] own violent proceedings ...

> Altogether Cyril presents a character not only unamiable, but singularly deficient in the graces of the Christian life...[30]

If the reader prefers the judgment of one of Cyril's contemporaries, we have this:

> His death made those who survived him joyful; but it grieved most probably the dead; and there is cause to fear, lest, finding his presence too troublesome, they should send him back to us... May it come to pass, by your prayers, that he may obtain mercy and forgiveness, and that the immeasurable grace of God may prevail over his wickedness![31]

So said Theodoret (*c.* 386–458), who wrote an ecclesiastical history from 322 to 427, independent of that written by Socrates Scholasticus.

The death of Hypatia in AD 415 marks an important place in the history of thought, which we may rank with Giordano Bruno's being burned at the stake (1600), Galileo's problems with the Inquisition (from 1616), and the purge of Soviet astronomers in 1936–7.[32] It was a time of scoundrels, and, as often happens, the guilty are praised in their day, rather than condemned. Cyril of Alexandria was made a saint.[33]

Not all the Church officials at that time had the attitude of Cyril and his uncle. One of these notable exceptions was Synesius of Cyrene (*c.* 373–414), who became bishop of Ptolemais (in Libya) in 410 and was confirmed in that office by none other than Theophilus. In great contrast to the times, Synesius upheld the individual's right to arrive at personal philosophic conclusions.[34] Correspondence between Synesius and Hypatia (which has survived) testifies to their mutual interests in science and Neoplatonism.

In spite of the general message put out by the clergy, scholarship for its own sake (i.e. without 'proper' religious relevance) was not dead. There were still mentors and students, books and commentaries, seminars and debates. Proclus (410–85), born in Constantinople and educated in Alexandria, became head of the Platonic Academy in Athens about 450. There he actively upheld the tradition of paganism (and survived the displeasure of the Church). He wrote an introduction to Ptolemaic astronomy (which has survived) that advocated the notion that there was no vacant space between the celestial spheres, a very popular idea until the time of Kepler. Among the pupils of Proclus were Asclepiodotos of Alexandria, Marinos of Sichem (who succeeded Proclus as head of the Academy in 485), and Ammonios, son of Hermias, who flourished in Alexandria and taught Damascios, Simplicios, and Joannes Philoponos.

In 529, by order of the Byzantine emperor Justinian, the Athenian Academy was closed. We should be saddened by its demise, but we should also celebrate that it lasted almost 900 years. In Alexandria, there is no single marking point of the 'end of scholarship', unless we take it to be the death of Hypatia and conclude, in retrospect, that scholarship died out slowly but surely after that.

The most significant event of the seventh century was the rise of Islam, which was widely disseminated in the years immediately following the death of Mohammed in 632. Under the general 'Amr ibn-el-'Ās, the Moslems battled the Greeks at Alexandria. 'Amr blockaded the city for 14 months, and the city finally succumbed in late 641. Concerning the destruction of the library at Alexandria, we have the following report of Bar-hebraeus (also known as Abu'l-Faraj or Abulfaragius), a Christian writer of thirteenth century Syria:

> John the Grammarian, a famous Peripatetic philosopher, being in Alexandria at the time of its capture, and in high favor with 'Amr, begged that we would give him the royal library. 'Amr told him that it was not in his power to grant such a request, but promised to write to the caliph for his consent. Omar, on hearing the request of his general, is said to have replied that if those books contained the same doctrine with the Koran, they could be of no use, since the Koran contained all necessary truths; but if they contained anything contrary to that book, they ought to be destroyed; and therefore, whatever their contents were, he ordered them to be burnt. Pursuant to this order, they were distributed among the public baths, of which there was a large number [4000] in the city, where, for six months, they served to supply the fires.[35]

In view of the fact that Abulfaragius wrote six centuries after the events in question, while Joannes Philoponos and his contemporaries have left us nothing about the survival *or* destruction of the library in their own day, the reader may decide for himself if the above-quoted account should be taken seriously. Gibbon does not believe that the library met its demise in the manner stated above; he believes that the account is inconsistent with the characters of both 'Amr and the Caliph.[36] As a result of the events in Alexandria from 390 to 641, it is entirely likely that there was little of the once-great library left to be burned.

It cannot be denied, however, that within a few years of the Arab conquest of Egypt, the character of Alexandria had radically changed. No longer was it the principal center of Greek commerce. No longer were there the conditions requisite for scholarship in the manner of the Athenian Academy or Alexandrian Museum. And, as the epitome of ancient astronomy is to be found in the accomplishments of the Greeks, with the end of the Greek institutions of higher learning, we come to the end of progress of ancient astronomy.

2

Astronomical capitals of the Moslem world

> It was He that gave the sun his brightness and the moon her light,
> ordaining her phases that you may learn to compute the seasons
> and the years ... It is He that has created for you the stars, so that
> they may guide you in the darkness of land and sea.
> *Koran*, 10:5, 6:97.

It is a common misconception that astronomical research fell into a dazed slumber following Ptolemy, not to reawaken until the time of Copernicus. I have briefly sketched in the previous chapter the efforts on the part of various Greeks in preserving their astronomical science. These efforts continued up to the time of the conquest of Egypt by the Arabs, who were not the book-burning fanatics that some have made them out to be. Those who think that these Arabs made no contributions of their own have not investigated the subject.

The principal astronomical centers of the Middle Ages were in the more exotic lands of the Middle East (greater Persia) and south central Asia (Turkestan). The principal astronomers were Moslems, Jews, and some Christians, but what they had in common was that they wrote in Arabic. This was the principal language of astronomy of the ninth through the eleventh centuries, just as English is today.

There were four principal centers of astronomy research in the Middle East and central Asia from the ninth to the fifteenth centuries. The first was established in Baghdad by 'Abdallāh al-Ma'mūn (786–833), the seventh Abbasid caliph (from 813 to 833). A great patron of philosophy and astronomy, al-Ma'mūn organized a scientific academy known as the House of Wisdom (Bayt al-ḥikma), which was the first such institution founded after the demise of the Alexandrian Museum. He built an observatory at Baghdad and another on the plain of Tadmor (in Syria). His astronomers determined the obliquity of the ecliptic to be 23° 33′. They carried out geodetic surveys and determined the diameter of the Earth to be 6500 miles. Al-Ma'mūn ordered the translation of scientific works into Arabic such as the translation of the *Almagest* by al-Ḥajjāj ibn Yūsuf in 829–30, which was

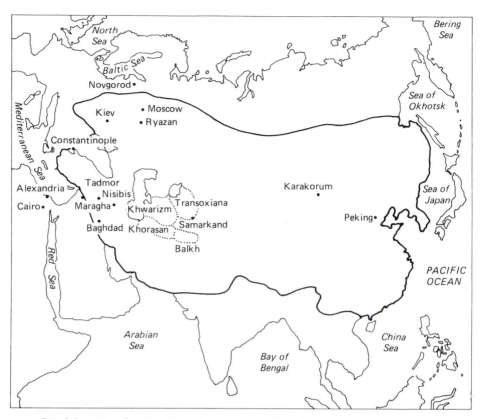

Fig. 2.1. Asia, showing important centers of medieval astronomy and also locations significant to the history of the Mongols. The dark outline is the extent of the Mongolian empire at its largest. After: *Encyclopaedia Britannica,* 11th edition, map in the article on Asia.

based on a Syriac version.* Many astronomers at Baghdad contributed observations for the so-called Tested Ma'mūnic Tables (832–3), which were tables of the motion of the planets based originally on Ptolemy's work, but updated on the basis of their contemporary observations. At that time mathematicians and astronomers were essentially one and the same. The most famous of these who worked under al-Ma'mūn's patronage was al-Khwārizmī, after whom we obtained the words *algorithm* (from his name) and *algebra*.[1]

Some of the astronomers who flourished in Baghdad at the time of al-Ma'mūn, along with their sons and grandsons, continued their work throughout the ninth century. Whereas, under the earlier Abbasid caliphs, the collaboration with non-Moslems was encouraged in an atmosphere

---

* The first translation of the *Almagest* into Arabic was carried out by Sahl al-Ṭabarī, a Jewish astronomer and physician, just prior to this. He was not associated with the Baghdad school.

conducive to the exchange and development of ideas, under the tenth Abbasid caliph Mutawakkil (847–61) these conditions drastically changed and the major scientific activity there rapidly diminished. What was determined to be orthodox Islam became of primary concern.

Cairo then became the center of astronomy research in the Islamic world. There, about 990, by order of the Fatimid caliph al-'Azīz the astronomer Ibn Yūnus (c. 950–1009) began to prepare improved astronomical tables, which were completed in 1007 under the next caliph al-Ḥākim. They are known as the Hakemite Tables (al-zīj al-kabīr al-Ḥākimī; the word *zij* means 'astronomical treatise' or 'astronomical tables', but a *zij* might also contain a catalogue of stars). Under al-Ḥākim another scientific academy was built, the Hall of Wisdom (dār al-ḥikma), which lasted from 1005 to 1171.[2]

It should be quite clear by this juncture that the Arabic astronomers were actively involved with observational astronomy and the theoretical aspects of astronomical theory (i.e. mathematical models). By this time they had become convinced that Ptolemy's value of 36″/year for precession was incorrect and that a larger value would be better for the production of accurate star catalogues. The famous astronomer al-Battānī (Albategnius; c. 858–929) obtained 54″.5/year,[3] while Ibn Yūnus found 51″.2/year,[4] remarkably close to the true value.

The next astronomical center of that time and region was at Maragha, but first we must speak of something entirely different, namely the most militarily supreme and ruthless conquerors the world has ever known: the Mongols. United under Genghis Khan (d. 1227), the Mongols soon swept in all directions and, within 15 years of their first leader's death, had put together the largest empire in all of history, at its largest comprising China, Russia, part of eastern Europe, parts of southern Asia and the Middle East. Along the way they destroyed countless cities and killed millions of people. As Matthew Paris, an English historian of the first half of the thirteenth century, describes them:

> They are inhuman and beastly, rather monsters than men, thirsting for and drinking blood, tearing and devouring the flesh of dogs and men, dressed in ox-hides, armed with plates of iron, short and stout, thickest, strong, invincible, indefatigable, their backs unprotected, their breasts covered with armor ... They are without human laws, more ferocious than lions or bears, ... drinking turbid or muddy water when blood fails them (as beverage). They have one edged swords and daggers, are wonderful archers, spare neither age, nor sex, nor condition ...[5]

Their typical method of war was to demonstrate their ruthlessness and destructiveness en route to some major city, then demand the city's surrender. *Whether or not the city gave up voluntarily*, it was usually destroyed and the inhabitants slaughtered. When they overran the Russian city of Ryazan in 1237 under the command of Batu Khan (a grandson of Genghis Khan)

> [o]n the dawn of the sixth day the pagan warriors began to storm the city, some with firebrands, some with battering rams, and others with countless scaling ladders for ascending the walls of the city. And they took the city of Ryazan on the 21st day of December. And the Tatars [i.e. those from hell] came to the Cathedral of the Assumption of the Blessed Virgin, and they cut to pieces the Great Princess Agrippina, her daughters-in-law, and other princesses. They burned to death the bishops and the priests and put the torch to the holy church . . . And the churches of God were destroyed, and much blood was spilled on the holy altars. And not one man remained alive in the city. All were dead. All had drunk the same bitter cup to the dregs. And there was not even anyone to mourn the dead.[6]

Hulagu (d. 1265), another grandson of Genghis Khan, and brother to Kublai Khan, was sent to quell disturbances in Persia organized by the Ismāʿīlī rulers there (also known as the Assassins). In 1256 Hulagu dispelled the troubles in typical Mongol fashion by obtaining submission from the enemy, then killing them all anyway. In 1258 the Mongols under Hulagu sacked Baghdad, reportedly killing 800 000 people.[7] He established in Persia the reign of the Ilkhans, which lasted effectively until 1335.[8] The Mongol ruler in Persia (the Ilkhan) acknowledged the supremacy of the Khakhan (supreme Mongol leader) in Karakorum,* but in essence functioned independently.

It is surprising to learn that within the Mongol tradition of butchery there were any intellectual accomplishments at all. But Hulagu 'freed' the astronomer Naṣir al-Dīn al-Ṭūsi (1201–74) from the Ismāʿīlīs and let him supervise the construction of an observatory at Maragha in Azerbaijan (now part of modern-day Iran). The observatory was begun in 1259. It covered a space 380 by 150 yards, and the foundation walls were almost 5 feet thick. Thus, it acted as a fortress as well.[9] The observatory contained many instruments in the Greek manner (armillary spheres, quadrants, parallactic rulers) and a great quadrant of radius 4.3 m.[10] The astronomers there had at their disposal a library of 400 000 manuscripts. Of Naṣr al-Dīn al Ṭūsi's 150 extant works, the best known is the *Zīj-i īlkhāni* or Ilkhanic Tables,

---

* At the start of the Yüan Dynasty in China (1260–1368), Kublai Khan moved the Mongol capital to Peking, which the Mongols called Khanbalik.

completed in 1272. These were originally written in Persian but were later translated into Arabic and Latin. Naṣr al-Dīn's *Tadhkira* (or *Treasure of Astronomy*) explained some of the major shortcomings of Ptolemaic astronomy and proposed some novel improvements (without, however, suggesting heliocentricism). These ideas probably reached Copernicus through Byzantine intermediaries.[11]

The Maragha observatory lasted perhaps 50 years. By the time Naṣr al-Dīn's collaborators and students had died out, so had that phase of astronomy. Persia and Turkestan were soon subjected to another reign of terror as extensive as the first wave of Mongolian sieges, this time organized by Timur the Lame (Tamerlane; 1336–1405). His father Tārāghāi, great-grandson of Karachar Nevian (a minister of Genghis Khan's son Jagatai), was head of the tribe of Berlas and his clan's first convert to Islam. Though Timur's father instilled in him a reverence for the Koran, Timur's life would be that of a conqueror motivated, it would seem, by the pure exhiliration of destruction. By 1370 when Timur had been proclaimed sovereign at Balkh (in northern Afghanistan) he was a powerful warrior chieftain. He soon made his capital at Samarkand (in the modern-day Uzbek Soviet Socialist Republic) and undertook his extensive campaigns, eventually counting among his dominions eastern Russia, part of Asia Minor, Armenia, Georgia, Syria, Iraq, Persia, Khorasan, Khwarizm, Transoxiana, Afghanistan and northern India. Not content with this, he sought to re-establish Mongol control of China, which had been ruled by the Mongols during the Yüan Dynasty. However, he died on the way to China. Timur was eventually buried in an elaborate jade tomb in Samarkand by his grandson Ulugh Beg.[12]

Ulugh Beg (or Ulug Beg, meaning 'great prince', a title which replaced his original name – Muhammed Taragai) was born in 1394 in the city of Sultania during a campaign of his fearsome grandfather. Ulugh Beg's father Shāh Rukh (named after a move in chess – check-rook) succeeded Timur in 1405 and a few years later made Ulugh Beg viceroy of the province of Transoxiana (Maverannakhr) between the River Oxus (Amu Darya) and the River Jaxartes (Syr Darya), the principal city of which was Samarkand. In contrast to his grandfather and father, Ulugh Beg's life was primarily devoted to scholarship, in particular, astronomy.[13–16] Just as Hulagu had commandeered various astronomers to build the observatory at Maragha, so had Timur and Shāh Rukh brought notable intellects to Samarkand. We have it on the authority of one Ghiyāth al-Dīn al-Kāshī (d. 1429) that systematic scientific work began there about 1408–10.[17]

Early in life Ulugh Beg had visited the site of the Maragha observatory, which made an impression on him.[18] About 1420, under Ulugh Beg's

Fig. 2.2. Ulugh Beg (1394–1449), Mongolian–Turkish ruler and astronomer. From Sheglov (Ref. 16).

patronage, a major observatory was constructed in Samarkand as part of a *madrasa*, or institution of higher learning. In contrast to other patrons of science such as al-Ma'mūn and al-Ḥākim, Ulugh Beg was a skilled mathematician and astronomer in his own right, having been taught by Qāḍī Zāda al-Rūmī (*c.* 1364–1436), a fellow Turk. Regarding Ulugh Beg's level of expertise, al-Kāshī tells us:

> His Majesty has great skill in the branches of mathematics. His accomplishment in these matters has reached such a degree that one day, while riding, he wished to find out to what day of the solar year a certain date would correspond which was known to be a Monday of the month of *Rajab* [the seventh month of the Islamic calendar] in the year eight hundred and eighteen [1415–16] and falling between the tenth and the fifteenth of that month. On the basis of these data he derived the longitude of the sun to a fraction of two minutes by mental calculation while riding on horseback, and when he got down he asked this servant [to check his result] ...

> Suffice it here to say that he is thoroughly skilled in this science [i.e. mental calculation], that he is proficient in giving proofs to astronomical

procedures, that he derives the rules as is due, and that he lectures on the *Tadhkira* [of Naṣr al-Dīn al-Ṭūsī] and the *Tuhfa* [of Quṭb al-Dīn al-Shīrāzī (1236–1311), a student of Naṣr al-Dīn al-Ṭūsī] in such a manner that the slightest addition to it cannot be imagined.[19]

The first director of Ulugh Beg's observatory was al-Kāshī, who was followed by Qāḍī Zāda. In addition to these men and their patron, observations were made by 'Alī ibn Muhammed al-Qūshchī (d. 1474). Al-Kāshī tells us that 'the most preeminent among the scientists are at the present [*c*. 1420] gathered together in Samarqand. Professors lecturing on all the sciences are numerous, and the majority among them is occupied with the mathematical sciences'.[20] He says that there are sixty to seventy such persons among them who know mathematics.[21] Furthermore,

> [a]s to the inquiry of those who ask why observations are not completed in one year but require ten or fifteen years, the situation is such that there are certain conditions suited to the determination of matters pertaining to the planets, and it is necessary to observe them when these conditions obtain. It is necessary, e.g. to have two eclipses in both of which the eclipsed parts are equal and to the same side, and both these eclipses have to take place near the same node. Likewise, another pair of eclipses conforming to other specifications is needed, and still other cases of a similar nature are required. It is necessary to observe Mercury at a time when it is at its maximum morning elongation and once at its maximum evening elongation, with the addition of certain other conditions, and a similar situation exists for the other planets.
>
> Now, all these circumstances do not obtain within a single year, so that observations cannot be made in one year. It is necessary to wait until the required circumstances obtain, and then if there is cloud at the awaited time, the opportunity will be lost and gone for another year or two until the like of it occurs once more. In this manner there is need for ten or fifteen years.[22]

The observations made at Samarkand were carried out over the years 1420 to 1437. The largest instrument there, and the largest meridian instrument ever built (then or since) was the so-called Fakhri sextant (Figs 2.3, 2.4), which was reported to be as large in radius as the height of the dome of the church of St. Sophia at Constantinople. After the demise of Ulugh Beg's observatory it was thought that this instrument had been destroyed but, on the basis of excavations carried out from 1908 to 1948,[23–5] the instrument was rediscovered and found to be a sextant of radius 40 m and consisting of two stone walls placed in the plane of the meridian and separated by a distance of 50.5 cm. It was so large that it could be used to make observations as accurate as 2–5 *seconds* of arc. The instrument was

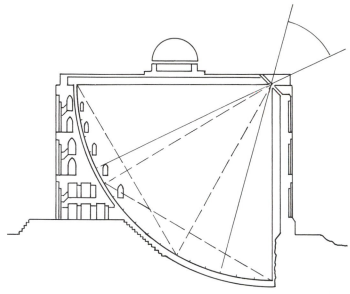

Fig. 2.3. A schematic diagram of the building for Ulugh Beg's colossal 40 m Fakhri sextant. A Fakhri sextant is a 60° arc placed in the plane of the meridian. It was invented by al-Khujandi in the second half of the tenth century. From Hoyle, Fred, *Astronomy*, London: Rathbone, 1962, p. 95.

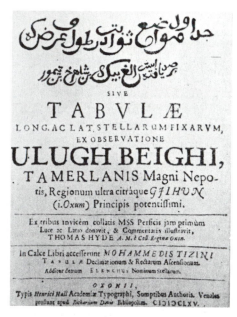

Fig. 2.4. (*a*) The Fakhri sextant at Ulugh Beg's observatory in Samarkand (modern excavation). From Marx, Siegfried, and Pfau, Werner, *Observatories of the World*, New York: Van Nostrand Reinhold, 1982, Plate 1, p. 9. (*b*) The title page of the catalogue of Ulugh Beg, published by Thomas Hyde (1665).

used mainly for determining the height of the Sun on the meridian, from which one can derive such parameters as the latitude, the obliquity of the ecliptic, the point of the vernal equinox, the length of the tropical year, etc. The astronomers there derived a value of $\varepsilon = 23° 30' 17''$ for the obliquity of the ecliptic (only 32" in error). Their value for the latitude of the observatory was 39° 37'.4 (which was 3'.2 in error). It is possible that meridian observations of the Moon and planets were also made with the Fakhri sextant, but it is doubtful that it was used for observations of fixed stars.[26] In addition to the large sextant, the Samarkand observatory contained armillary spheres, a triquetrum (Ptolemaic rulers), an azimuthal quadrant, astrolabes, and a *shāmila* (which served as an astrolabe and a quadrant).[27] These were used for extrameridional observations.

The *magnum opus* of the Samarkand observatory is known as the *Zij* of Ulugh Beg or the *Zij-i Gurgani* (after Guragon, the title of Genghis Khan's son-in-law, which was also used by Ulugh Beg). It was originally written in the Tadzhik language. It contains a complete survey of the astronomy of the fifteenth century, much like Ptolemy's *Almagest* summarizes the astronomical knowledge of the second century. Ulugh Beg's *Zij* contains information on time reckoning, calendars, practical astronomy (including trigonometry), the motions of the Sun, Moon, and planets, and the positions of the stars. As indicative of the accuracy of the tables, the mean yearly motions of the planets were known to better than 5 arc seconds but, if their time line stretched back to include the observations by the Babylonians (contained in the *Almagest*, which was available), this might not have been so hard to achieve. Still, Ulugh Beg's value of precession (51".4/year) is quite close to the actual value for that time (50".2/year).[28]

The greatest achievement of fifteenth century observational astronomy, embodied in Ulugh Beg's *Zij*, was the catalogue of 1018 stars, giving their ecliptic coordinates for the epoch 1437. It was the first systematic determination of the positions of all the stars visible with the naked eye (from that latitude) since the time of Hipparchus seventeen centuries before.

Tens of copies of Ulugh Beg's *Zij*, many of which are in Persian or Arabic, are to be found in libraries in Europe and Asia, but while the *Zij* was a product of the first half of the fifteenth century, it only became known in Europe in 1648 owing to the efforts of John Greaves (1602–52), the Savilian professor of astronomy at Oxford. In that year he published the latitudes and longitudes of about a hundred of the principal stars in Ulugh Beg's catalogue, and also, in a separate edition, a translation of the geographical tables from the *Zij*. Two years later Greaves published a translation of the first part of the *Zij* dealing with chronology. Greaves died, however, before he could fulfill his ambition of publishing the complete star catalogue. Not long after this the curator of the Bodleian Library at Oxford, Thomas Hyde (1636–1703) edited and published a Tadzhik and Latin edition of the complete star catalogue (1665). Other works pertaining to Ulugh Beg's *Zij* are the *Prodromus Astronomiae* of Johannes Hevelius (1690), John Flamsteed's *Historia Coelestis Britannica* (1725; see chapter 4), and L. P. E. A. Sédillot's *Prolégomènes des Tables Astronomiques d'Oloug Beg* (1847–53). Other editions of the star catalogue include those published by G. Sharpe (1767), Francis Baily (1843), and Edward Ball Knobel (1917).[29] Sarton considers Ulugh Beg's work to rank as one of the three most significant masterpieces of Muslim observational astronomy, the other two being the Book of the Fixed Stars by 'Abd al-Raḥmān al-Ṣūfī (903–86), and the Hakemite Tables of Ibn Yūnus.[30]

As to how the observations were carried out at Samarkand, al-Kāshī tells us that only one or two people needed to be involved with the actual observations at any time,[31] while a larger number of people were involved with the discussion of the results, much like a modern peer-review procedure:

> [Ulugh Beg] has allowed that in scientific questions there should be no agreeing until the matter is thoroughly understood and that people should not pretend to understand in order to be pleasing. Occasionally, when someone assented to His Majesty's view out of submission to his authority, His Majesty reprimanded him by saying 'you are imputing ignorance to me'. He also poses a false question, so that if anyone accepts it out of politeness he will reintroduce the matter and put the man to shame.[32]

The observations for Ulugh Beg's *Zij* were primarily completed by 1437, and it was probably a few years after that that the data were reduced and the astronomical tables completed. Much in contrast to a modern observatory, ongoing observations and further refinement of the mathematical elements were not regarded as quite so important. Meanwhile, Ulugh Beg's father died in 1447 giving him, as the successor, greater administrative responsibilities as the khan of a large region. But his rule was of short duration. In 1449 he was killed by an assassin hired by his son Abdul Latif.[33]

> In 1941 an expedition under the leadership of T. N. Kari-Niazov discovered the tomb of Ulugh Beg in the mausoleum of Tamerlane in Samarkand. In contrast with the Islamic custom of burying the dead only in a shroud, Ulugh Beg lay fully clothed in a sarcophagus, in agreement with the prescription of the *shariat*: a man who died as a *shakhid* (martyr) had to be buried in his clothes. On the skeleton, traces of his violent death are clear: the third cervical vertebra was severed by a

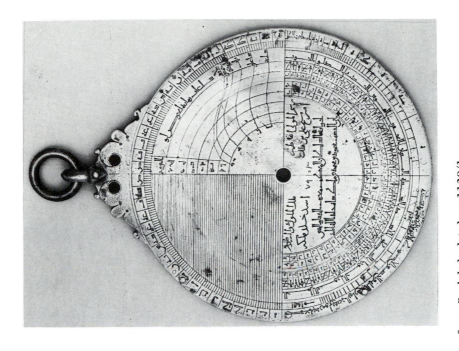

Fig. 2.5. The two sides of an astrolabe from Baghdad, dated AD 1130/1. Photographs courtesy Adler Planetarium, Chicago.

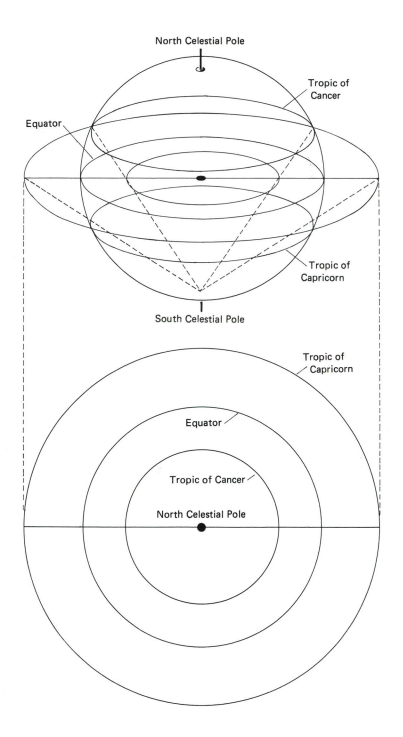

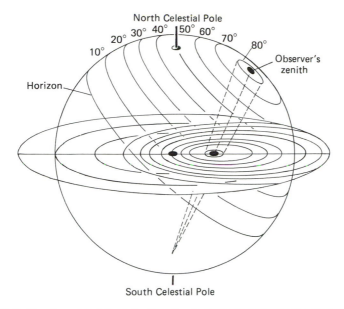

Fig. 2.6. For stereographic projection the sky is projected onto the plane parallel to the celestial equator. It has the property that all circles on the sky (whether great or small circles) map as circles onto the plane of projection. This example is for north latitude 40°. From North, J.D. 'The astrolabe', *Scientific American*, January 1974, pp. 99, 100.

sharp instrument in such a way that the main portion of the body and an arc of that vertebra were cut off cleanly; the blow, struck from the left, also cut through the right corner of the lower jaw and its lower edge.[34]

It was no consolation that Abdul Latif was himself murdered a year later. The death of Ulugh Beg effectively put an end to the flourishing scientific activity in Samarkand.

The astronomical instrumentation of the Arabs surpassed that of the Greeks in a number of ways.[35] There had been improvements over time in the techniques of metal working, and metal is easier to fashion than stone and does not warp like wood. Larger instruments were built which allowed the measurement of positions on the sky with better angular resolution. The largest of these, already alluded to, was the Fakhri sextant at Samarkand, an instrument invented by al-Khujandī in the second half of the tenth century.[36] Finally, the Arabs must have recognized that with *permanently mounted* instruments it was possible to achieve more accurate results. Anyone who has a portable instrument realizes that if it is eventually situated in a permanent spot, such things as the levelling of the ground and laying down the meridian line are not problems for each observing session.

One such instrument we have not yet discussed is the *astrolabe*, from the

Greek ἄστρον (star) and λαβεῖν (to take, to find) – a star finder. There were planar astrolabes and there were spherical astrolabes, which were not the same as armillary spheres. The spherical astrolabe was like a globe and was used to demonstrate the positions of stars in the sky along with the systems of coordinates, while the armillary sphere was used for actually sighting celestial objects and reading off their coordinates. (In the *Almagest* Ptolemy called the armillary sphere an *Astrolabon organon*, or star-finding instrument.) Only one spherical astrolabe is still extant. It was built in the fifteenth century.[37]

The term astrolabe, as it has been used since the rise of Arabic astronomy, refers to the plane, or planispheric, astrolabe, which is based on the principal of *stereographic projection*. According to Synesius of Cyrene (the friend of Hypatia) it was Hipparchus who discovered that the celestial sphere could be projected from the South Pole onto a plane parallel to the plane of the celestial equator, with the property that circles on the celestial sphere remained circles in the plane of projection. (The celestial meridian, which is a great circle in the sky, becomes a straight line on the astrolabe because it is projected on edge.) A rotatable star grid could be made for the astrolabe which preserved the distances between the stars as the grid was turned on the astrolabe according to the apparent daily rotation of the sky. One could draw on the plane of projection the loci of equal elevation angles above the horizon (*almucantars*) and the loci of equal azimuths, for observations at a given geographical latitude. These latitude-dependent graphs are called *climates*, in an obvious extension of meaning of the word (see Fig. 2.7). Then, by observing the elevation angle of a star or the Sun with some kind of sighting device, one could use the plane astrolabe as a calculating device to determine the local time, either by the stars (sidereal time) or by the Sun (solar time), since it was possible to calculate the position of the Sun on the ecliptic for any given day of the year. The star grid took the form of a spider's web-like filigree with the positions of labelled stars corresponding to the points in the pattern (see Fig. 2.5). The grid is known as a *rete* after the Latin word for net. A typical rete might represent the positions of 15–50 stars. Only several of the brightest stars are really necessary because the observation of *any* single star above the horizon could be used to fix the time, and it is less difficult to sight a brighter star than a fainter star. The rete also contained a projection of the ecliptic, making it possible to rotate the rete until the point on the ecliptic corresponding to the position of the Sun matched the observed elevation angle of the Sun in the eastern or western half of the sky.

The sighting device, usually to be found on the back of the astrolabe, is called the *alidade*, from the Arabic *al 'idadah*, meaning turning radius. It is a

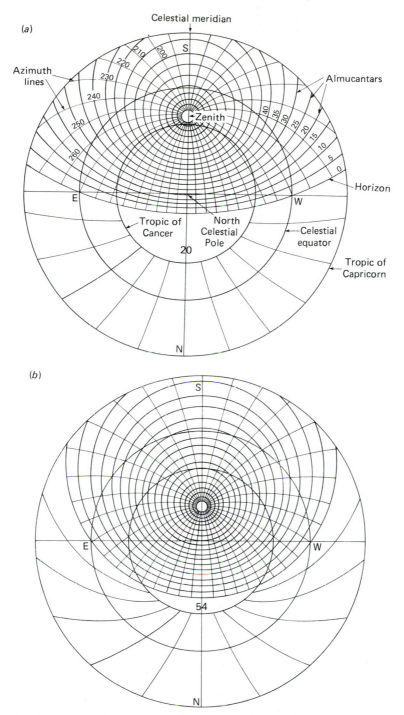

Fig. 2.7. (a) Example of a climate for an astrolabe for a latitude of 20°. As a traveler moved north or south he would use a climate appropriate for his latitude. The rete, or overlay for the positions of the stars, which was placed over the climate, only became out of date after many decades when the star coordinates had changed significantly due to the effect of the precession of the equinoxes. (b) Example of a climate for a latitude of 54°.

straight edge which points to degree markings around the edge while it turns
about the center of the astrolabe. The alidade has a sight at each end made up
of a flat piece of metal with a small hole (about 2 mm) drilled through. One
hangs the instrument from a finger of one hand and rotates the alidade until
one can sight a known star through the two holes. In the case of the Sun one
would cast the round image of the Sun through the top hole onto the lower
sighting hole. One then simply reads off the elevation angle from the outside
edge of the back of the astrolabe. Then, on the front of the astrolabe, one
rotates the rete to place the star or Sun at the correct elevation angle in the
eastern or western half of the climate. It is then possible to determine the
elevation angles and azimuths of all the other stars represented in the rete
(without any calculating whatsoever), determine the local sidereal time (the
right ascension corresponding to the meridian), or to determine the local
solar time. Then, using one's known geographical longitude and a graph of
the *equation of time* (the difference between mean and apparent solar time) it
is possible to derive the modern standard time (e.g. Greenwich Mean Time).
With an astrolabe it is easily possible to determine the time to within 5 min
but, because of precession of the equinoxes, an astrolabe's rete eventually
becomes out of date. This gives an advantage to the historian, however,
because one can determine from the astrolabe itself not only for what latitude
it was intended, but also approximately what year it was made.[38]

As mentioned above, Hipparchus was familiar with stereographic projection, and may have invented it.[39] The details of this geometrical basis of the
planispheric astrolabe were laid down in Ptolemy's *Planisphaerium*, the
original Greek version of which has not survived; but an Arabic translation
of it exists, along with a Latin translation of that, and a German translation of
that (!).[40] Though Ptolemy knew of the principles of constructing astrolabes,
it is not known if he actually made them. Theon of Alexandria, the
commentator on the *Almagest* and father of Hypatia, wrote a treatise on the
astrolabe (*c.* 375). The earliest treatise on the construction and use of the
astrolabe that has come down to us is by Joannes Philoponos (*c.* 530),[41]
while Theon's treatise is largely contained in another, similar work, that of
Bishop Severus Sebokht (*c.* 650) of Nisibis in Syria.[42] Knowledge of the
astrolabe then passed into the hands of the Arabs, who made it the most
popular astronomical instrument for a thousand years. To name all the
medieval astrolabe makers or writers on the subject would be tedious,
but such a list would include many of the major Arabic astronomers
such as al-Bīrūnī (973–1048), continental Europeans such as Gerbert
(*c.* 945–1003; later Pope Sylvester II) and Hermann the Lame of
Reichenau (1013–54), and the famous Englishman and author of the

Canterbury Tales, Geoffrey Chaucer, who wrote his *Treatise on the Astrolabe* in 1391.[43]

Medieval astrolabe makers could earn a good living, as their instruments were both beautiful and functional.[44] They were used by astronomers and surveyors, and were also used in the casting of horoscopes (perhaps the major reason for their popularity). Astrolabes were used in Europe until the end of the seventeenth century, when they were replaced by instruments containing telescopes having eyepieces with reticles. In the Islamic world they were still used in the nineteenth century.

3

Tycho Brahe's observatory at Hven

The reformation of astronomy should proceed not by the authority
of men but from precise observations and deductions from them.
Tycho Brahe

The astronomy of the ancients, with its Arabic improvements, reached
Europe via Moorish Spain, where it flourished first at Cordova (tenth
century), then at Toledo (eleventh century).[1] The astronomer al-Zarqālī
(c. 1029–87) composed the *Toledan Tables*, which were tables of the predicted
positions of the planets, with appropriate explanations, based on observations by himself and other Muslim and Jewish astronomers in Toledo. These
were updated under the patronage of the Christian king Alfonso X of Castile
in 1272; the *Alphonsine Tables* continued to be widely used and were
reprinted as late as 1641.[2] The *Toledan Tables* were translated into Latin by
Gerard of Cremona (1114–87), whose love of the *Almagest* led him to study
Arabic, and who became one of the most important and prolific translators of
all time. (Particularly at this time, the transmission of known science was as
important as the discovery of new science.) The first Latin edition of the
Almagest (Venice, 1515) was Gerard's, of which Copernicus obtained a
copy.[3]

Fifteenth century developments in Central Europe were epitomized by two
men: Georg Peurbach (1423–61), whose *Epitoma Almagesti Ptolemaei* was a
translation with commentaries of the Greek (not Arabic) text of the *Almagest*;
and Peurbach's protégé and collaborator, Johann Müller (1436–76), known
as Regiomontanus, the most important astronomer of the fifteenth century,
who completed Peurbach's work and was the first publisher of mathematical
and astronomical literature. Copernicus was probably acquainted with
every publication written, edited, or printed by Regiomontanus.[4] The
Nuremberg home of Regiomontanus was also perhaps the first observatory
in Europe worthy of the name.[5,6] This brings up the interesting question of
what constitutes an observatory. Is it a place where observations are made
(with a highly portable instrument or perhaps with no instrument at all), or
is it a place with 'permanently' mounted instruments and a systematic
research program?

Fig. 3.1. Nicholas Copernicus (1473–1543), the founder of modern astronomy.
From Hoyle, Fred, *Astronomy*, London: Rathbone, 1962, p. 97.

This is not the place to discuss in detail the founder of modern astronomy, Nicholas Copernicus (1473–1543; Fig. 3.1), whose revolutionary treatise *De revolutionibus orbium coelestium* (1543) must be ranked with Newton's *Principia* and Einstein's Theory of Relativity as one of the three most significant works in astronomy and physics since ancient times. Copernicus was not the first to contemplate a heliocentric solar system. He himself knew of the ideas of Aristarchus of Samos (*c.* 280 BC). But the ancients could not derive a convincing heliocentric theory or demonstrate that the Earth moved or rotated. Copernicus' great book was not a wholesale replacement of the Ptolemaic theory either – while Ptolemy required 40 epicycles, Copernicus retained the idea of circular motion and needed 34 epicycles (7 for Mercury, 5 each for Venus, Mars, Jupiter, and Saturn, 4 for the Moon, and 3 for the Earth).[7] The time was ripe for a reappraisal of the Ptolemaic theory, and to Copernicus is attributed the credit for rearranging the cosmos.

Copernicus was no great observer, only 27 observations being credited to

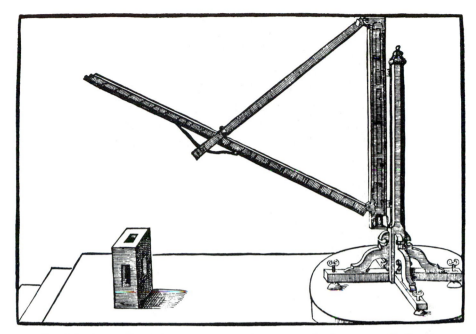

Fig. 3.2. A parallactic instrument, or triquetrum, also known as Ptolemaic rulers. Tycho Brahe built this one based on a model of Copernicus he had acquired. Photograph courtesy Archenhold-Sternwarte, Berlin.

him.[8] He owned only three instruments:[9] a quadrant mounted in the plane of the meridian for measuring the Sun's elevation; a wooden parallactical instrument (like that in Fig. 3.2), which passed into the hands of Tycho Brahe;[10] and a radius astronomicus, or Jacob's staff, which is made up of a calibrated staff with a cross piece that slides along it, for measurements of large angular distances.

Though Copernicus made plans for a permanent observing place for his quadrant and gives details for its accurate alignment, his best-laid schemes were minor compared to the efforts of Wilhelm IV, the Landgrave of Hesse-Cassel (1532–92). The Landgrave constructed an observatory very worthy of the name and was the first to situate an instrument in a rotatable dome.[11] Wilhelm observed systematically from 1561 to 1567 with a number of instruments, some of which were made of brass such as an azimuthal (rotatable) quadrant, and planned to make a more accurate star catalogue. His was 'the first observatory in central Europe which was placed on a firmer footing and whose activity was carried out with zeal for a number of years in a particular direction'.[12] His star catalogue was not finished but, under his direction, the transition of the instruments from wooden ones to metal ones was on the whole completed. Though his azimuthal quadrant was an

Fig. 3.3. Tycho Brahe (1546–1601). He lost part of his nose in a duel with another Danish nobleman in 1566. The prosthetic piece was made of gold, silver and copper and is clearly evident in this picture. Photograph courtesy of Archenhold-Sternwarte, Berlin.

innovation, it led to the development of instruments similar to it which were to be *fixed* on the meridian.[13] (Until the development of the transit circle in the nineteenth century, astronomers would use two different instruments fixed on the meridian for the separate determination of right ascensions and declinations.)

The observatory built by Tycho Brahe on the Danish island of Hven was so magnificent that it can be regarded as the most elaborate astronomical undertaking of all time. The total cost of the buildings and instruments amounted to 75 000 *daler* (over several years), or about 30% of the annual income of the Danish government at that time.[14] Imagine what kind of observatory Denmark could build today for the equivalent of $5 billion.[15]

To backtrack a bit, Tycho Brahe (1546–1601) was born at Knutstorp (or Knudstrup) in the southern part of modern day Sweden (then part of Denmark). His interest in astronomy was sparked by the occurrence of a solar eclipse visible in Copenhagen in 1560. Three years later Tycho observed a conjunction of Jupiter and Saturn and was impressed to note that the time of closest approach was more accurately predicted by Erasmus Reinhold's *Prutenic Tables* (based on Copernicus) than by the *Alphonsine Tables* (based on Ptolemy). While the latter erred by a month, the fact that Copernican theory still led to an error of several days made the Landgrave of Hesse-Cassel and Tycho Brahe both resolve to improve the state of astronomical observations.

On 11 November 1572 Tycho made a startling discovery:

> Raising my eyes, as usual, during one of my walks, to the well-known vault of heaven, I observed, with indescribable astonishment, near the zenith, in Cassiopeia, a radiant fixed star, of a magnitude never before seen ... This new star I found to be without a tail, not surrounded by any nebula, and perfectly like all other fixed stars, with the exception that it scintillated more strongly than stars of the first magnitude ... Those gifted with keen sight could, when the air was clear, discern the new star in the daytime, and even at noon ... Its distances from the nearest stars of Cassiopeia, which, throughout the whole of the following year, I measured with great care, convinced me of its perfect immobility.[16]

Tycho's supernova remained visible for 17 months. For some of his observations Tycho used a wooden quadrant of radius 5.4 m, set up near Augsburg. Not only did the supernova stimulate the construction of an instrument large enough to provide observations with a resolution better than 1 minute of arc, but the observation that change took place beyond the sublunar sphere, in the starry realms, where Tycho correctly placed the object on the basis of its lack of motion and parallax, refuted the ancient notion of the 'incorruptibility of the heavens'.

In 1575 Tycho paid a visit to the Landgrave Wilhelm. They observed together. They discussed the construction of instruments and the requirements of astronomy. In 1576, possibly because of a recommendation by the Landgrave, King Frederick II of Denmark bestowed upon Tycho the 2000 acre island of Hven (or Hveen; Venusia; Ven; Fig. 3.4) in the Danish sound between Copenhagen and Elsinore, so that he might establish a permanent observatory and carry out astronomical observations with a thoroughness never before achieved. He was to receive all the rents and duties of 'his' subjects on the island. Tycho received a royal estate in Norway, a stipend allotted from the Roskilde Cathedral, and an annual grant from the king's privy purse of 500 *daler*. This gave him an annual income of some 2400 *daler*, almost 15 times the salary of a professor at Copenhagen at that time.[17] Though he was one of the first full-time scientists in Europe, he 'contributed to the widening gap between teaching and research that had originated with

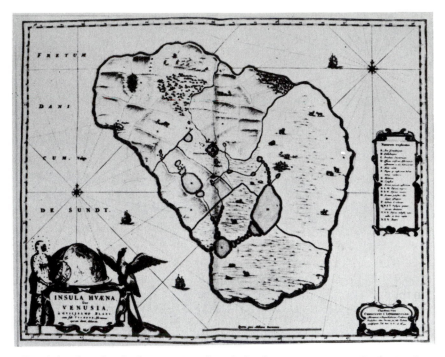

Fig. 3.4. The island of Hven, site of Tycho's observatory. From: Hoyle, Fred,
Astronomy, London: Rathbone, 1962, p. 109.

Regiomontanus at Nuremberg, had become apparent with Copernicus at
Frombork, and later was to assume rather disastrous proportions with the
great 17th-century academies'.[18] To put ourselves in Tycho's shoes,
however, if you could live like a king *and* make great progress in your chosen
field, would you choose to live the poorer life of a scholar, teach required
courses, and do less research? Indeed, Tycho lived like a king, ruled Hven like
one, and even entertained notable visitors such as James VI of Scotland (later
James I of England).

The construction of Uraniborg ('Castle of the Heavens') was begun in
1576 and completed in 1580 (Figs 3.5 & 3.6). It was built at the highest
point of the island, 160 feet above sea level. Though Tycho gives many
details of the buildings and the instruments in his *Astronomiae Instauratae
Mechanica* (Wandsbeck, 1598),[19] it is difficult to say exactly what size
everything was because of his system of units.[20]

The castle was built in the center of a square enclosure of walls. In 1868
H. L. d'Arrest measured the size of the square enclosure to be 254.7 Parisian
feet (82.8 m). In the *Mechanica* Tycho gives this as 300 feet. In Tychonian
units,

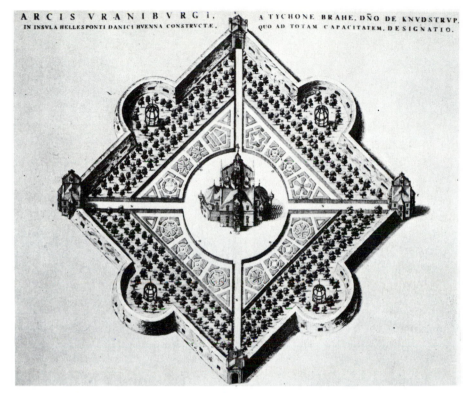

ARCIS VRANIBVRGI,
IN INSVLA HELLESPONTI DANICI HVENNA CONSTRVCTÆ.
A TYCHONE BRAHE, DÑO DE KNVDSTRVP,
QVO AD TOTAM CAPACITATEM, DESIGNATIO.

Fig. 3.5. Tycho Brahe's Uraniborg observatory. From Hoyle, Fred, *Astronomy*, London: Rathbone, 1962, p. 108.

the wall is twenty feet in breadth below, and is twenty-two feet high, and the inside diameter of the semicircular wall [see Fig. 3.5] measures ninety feet. The house in the centre, which also ... is of an exactly square form, on each side measures nearly sixty feet, the wall is forty-five feet high. But the round towers added in the south and the north are twenty-two feet in diameter, to which should be added ten feet for the galleries encircling them. The square portals on the eastern and western sides measure below fifteen feet in each direction. The complete height of the house itself to the Pegasus [weathervane] at the very top amounts to seventy-five feet. Further, the house itself in its whole extent is provided with a basement twelve feet deep.[21]

Below this there is even another level. Besides the principal observing locations, the castle had a dozen or so bedrooms, a dining room, a library, kitchen, chemical laboratory, and was equipped with running water.[22]

Table 3.1 lists the principal instruments used at Uraniborg. For the most part they are 1.5–3 m in size and are made of brass, iron, or steel. All the sighting was done by the naked eye, without magnification (the telescope

Fig. 3.6. Cross section of Uraniborg castle. Photograph courtesy of Archenhold-
Sternwarte, Berlin.

was not invented until 1609 or so). To improve the accuracy of the
measurements, Tycho devised special sights (Fig. 3.7) and used transversals
for the interpolation between markings on the arcs (Fig. 3.8). The largest
instruments, such as the mural quadrant (Fig. 3.9) allowed measurements
to be made with a resolution of 10 arc seconds. Instruments such as the
zodiacal armillary (Fig. 3.10) allowed the ecliptic coordinates of an object to
be measured. For direct measurement of right ascensions and declinations
Tycho also used equatorial armillaries; he was the first astronomer in Europe
to use the celestial equator, rather than the ecliptic, as a great circle for direct
reference.

Uraniborg was so overflowing with instruments after a few years that in
1584 Tycho had another observatory built about 100 feet to the south,
known as Stjerneborg ('Castle of the Stars'; Fig. 3.11). It contained some of
the most advanced instruments Tycho made (see Table 3.2), such as the
great equatorial armillary (Fig. 3.12), 7 cubits (about 2.9 m) in diameter,

Table 3.1 *Instruments used at Uraniborg*

Location	Instrument	Refs[a]	Size[b]	Comments
Larger southern observatory	Brass azimuthal quadrant	3/22	$r = 0.63$ m	1' resolution; brass lattice work
"	Brass sextant	4/23	$r = 1.67$ m	For measuring altitudes; cast brass plates; on wooden mounting
"	Azimuthal semicircle	8/27	$d = 2.5$ m	
"	Parallactic instrument	9/28	$l = 1.7$ m	Wooden; for measuring zenith distance of the Moon; patterned after Copernicus
Smaller southern observatory	Equatorial armillary	13/32	$r = 1.6$ m	Iron and brass
Larger northern observatory	Large brass parallactic instrument	10/29	$l = 3.5$ m	Gave altitudes and azimuths
"	Bipartite arc	15/34	$l = 1.7$ m	For measuring small angular distances; 0.6 m at top end; 1.2 m at bottom; required 2 observers
"	Iron sextant	17/36	$l = 1.3$ m	For measuring large angular distances
"	Steel quadrant in a square	18/37	$r = 2.0$ m	Used to observe 1572 supernova
"	Copernican parallactic instrument	9/—	$l = 1.7$ m	Wooden
Smaller northern observatory	Equatorial armillary	12/31	$d = 1.6$ m	Iron, brass, and wood; for measuring right ascensions and declinations
Southwest room on ground floor	Mural quadrant	5/24	$r = 2.0$ m	10" resolution; brass arc 5" wide, 2" thick; for measuring altitudes of stars and times of meridian transit

[a](Raeder *et al.* (1946) section number)/(Repsold (1908) Fig. number)
[b]Using Repsold's conversion factor of 1 cubit = 417 mm.

Fig. 3.7. Tycho's instrumental sights were specially designed to minimize errors. Photograph courtesy of Archenhold-Sternwarte, Berlin.

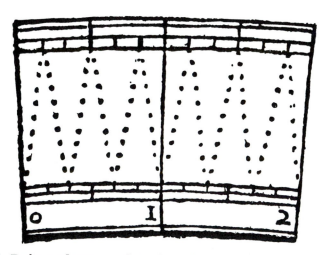

Fig. 3.8. Tycho used transversals to obtain the greatest resolution possible in making visual readings. Photograph courtesy of Archenhold-Sternwarte, Berlin.

Fig. 3.9. Tycho's mural quadrant. Photograph courtesy of Archenhold-Sternwarte, Berlin.

which was made of iron. On it a degree was so large that every minute was subdivided into four parts. It was primarily used for the determination of declinations. The noteworthy attribute of the Stjerneborg was that the instruments were situated in subterranean nooks in order to provide shelter from the wind. There were stairs leading down to the bases of the instruments for the convenience of the observer to sit or stand on. Three of the crypts had folding roofs, while two were provided with revolving domes.[23]

The instruments listed in Tables 3.1 and 3.2 do not exhaust Tycho's astronomical weaponry. Certainly one man could not use it all by himself,

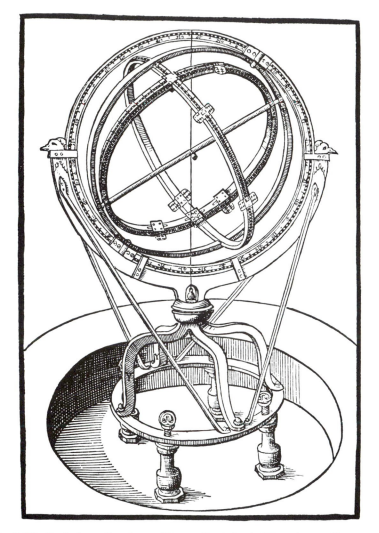

Fig. 3.10. Tycho's zodiacal armillary, situated at Stjerneborg. Photograph
courtesy of Archenhold-Sternwarte, Berlin.

and Tycho employed a number of observing assistants over the years, among
them Peter Jacobsøn Flemløs, Christian Severin (known as Long-
omontanus), Elias Olsen, Gellius Sascerides, Oddur Einarsson, Willem Blaeu,
and Paul Wittich.[24]

  'The organization of every modern observatory derives from Uraniborg,
though the instruments used there have been developed far beyond Brahe's
possibilities and even his conceptions.'[25] Tycho was the first astronomer ever
to describe his methods of observing in detail. The descriptions of his
instruments in the *Mechanica* give us a clear picture of the state of his

Table 3.2. *Instruments used at Stjerneborg*

Instrument	Refs[a]	Size[b]	Comments
Revolving azimuthal quadrant	6/25	$r = 2.5\,\text{m}$	15″ resolution; cast iron
Steel quadrant in a square	7/26	$r = 2.0\,\text{m}$	10″ resolution; steel with brass plated arc
Zodiacal armillary	11/30	$d = 1.25\,\text{m}$	1–2′ resolution; iron and brass
Great equatorial armillary	14/33	$d = 2.9\,\text{m}$	15″ resolution; iron; for measuring declinations
Four cubit sextant	16/35	$l = 1.7\,\text{m}$	For measuring large angular distances; had graduated brass arc

[a](Raeder *et al.* (1946) section number)/(Repsold (1908) Fig. number).
[b]Using Repsold's conversion factor of 1 cubit $= 417\,\text{mm}$.

astronomical capital.* It even had a printing press and a paper mill so that Tycho might disseminate his astronomical discoveries to the learned world.

After the observations of the 1572 supernova, Tycho's next project of note was his series of observations of the comet of 1577.[26] These were published in *De mundi aetherei recentioribus phaenomenis* (1588), which was printed at Tycho's own press at Uraniborg. Tycho demonstrated that the comet was not an exhalation of the Earth's atmosphere (as held by the Aristotelians) – it was at least three times as far away as the Moon. The orbit was shown to be an oval, countering the notion that motion must be circular. Tycho's observations of the nova of 1572 and the comet of 1577 challenged the notion of the Ptolemaic celestial spheres: the nova showed that the sphere for the stars was corruptible, and the comet passed through a number of the supposed spheres.[27] These thoughts led Tycho to formulate the Tychonic version of the solar system (Fig. 3.13): he placed the Earth at the center, the Moon orbited the Earth, while Mercury, Venus, Mars, Jupiter and Saturn orbited the Sun, and the Sun and its five planets orbited the fixed Earth. Though this has been regarded as a great leap backwards, it was based on observational evidence. Tycho's measurements (to an accuracy approaching 1′) exhibited no evidence of annual parallax of the stars, and he could not imagine that the stars were more than 3438 Earth–Sun distances away (the

* The only other comparable treatise on the outfitting of an observatory was W. Struve's 1845 *Description* of Pulkovo Observatory (Chapter 5), which served as a model for many other nineteenth century observatories.

Fig. 3.11. The Stjerneborg observatory at Hven, completed in 1584. From
Hoyle, Fred, *Astronomy*, London: Rathbone, 1962, p. 108.

cotangent of 1′ is 3438). He believed that the nearest stars were only just
beyond Saturn, with distances of about 14 000 Earth radii.[28] Even though
Tycho underestimated the distance to the Sun by more than a factor of 20, he
knew that the accuracy of his instruments implied that the stars were at least
4 million Earth radii distant if the Earth orbited the Sun. As the parallaxes are
less than 1″ in actuality, implying distances greater than 206 265
Earth–Sun distances (or Astronomical Units), the modern equivalent of
almost 5 billion Earth radii, it is no wonder that parallax determinations
were not demonstrated until long after Tycho (about 1838).

Tycho's primary task at Hven was to catalogue the positions of the stars
and make observations of the Moon, Sun, and planets in order to understand
their motions better, the very same task which would occupy the astronomers at Greenwich beginning a century later for their improvements of
the art of navigation (Chapter 4). Tycho did not pursue such practical means
per se but, because he knew he possessed the best collection of observational
instruments in the world, he knew that his results would stand as a
landmark in the progress of astronomy.

Manuscript copies of Tycho's star catalogue, containing about a thousand
stars, were circulated during Tycho's lifetime, but his star catalogue was not
published until after his death. In 1602 Johannes Kepler published the

Fig. 3.12. Tycho's large equatorial armillary, situated at Stjerneborg. Photograph courtesy of Archenhold-Sternwarte, Berlin.

Astronomiae instauratae progymnasmata, containing a reprint of Tycho's work on the nova of 1572, Tycho's lunar and solar theories, and a catalogue of 777 stars. This was increased to 1005 stars in a subsequent edition of the catalogue published at the end of Kepler's *Rudolphine Tables* (1627).[29] One can compare Tycho Brahe's star positions to those adopted in the *Alphonsine Tables* and those adopted by Ulugh Beg, Copernicus, the Landgrave of Hesse, and John Flamsteed by consulting Flamsteed's *Historia Coelestis Britannica* of 1725 (see Chapter 4).[30]

As mentioned previously, Tycho's most accurately inscribed instruments achieved a *resolution* of about $\frac{1}{6}$ of an arc minute. This does not mean that his

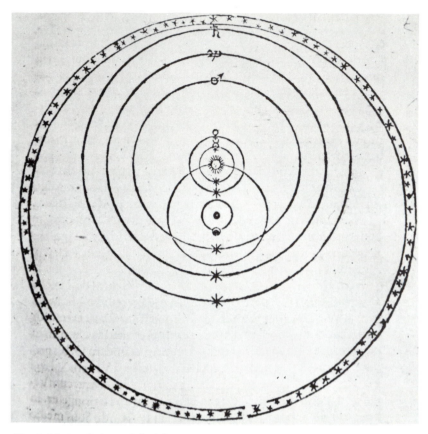

Fig. 3.13. Tycho's model of the arrangement of the solar system. The Earth is
situated at the center. The Moon orbits the Earth. Mercury, Venus, Mars, Jupiter,
and Saturn orbit the Sun, which in turn orbits the Earth. From Tycho Brahe, *De
mundi aetheri recentioribus phaenomenis*, 1588, reprinted by permission of the
Houghton Library, Harvard University.

observations can be trusted to that precision. There are several reasons:
(*a*) Though Tycho was one of the first to consider the effect of atmospheric
refraction, he believed the effect was negligible at elevation angles greater
than 20° (introducing possible errors of up to 3'), while at lower elevations
his model was just a first step.[31] (*b*) To convert relative coordinates of stars to
absolute coordinates (right ascension and declination) by means of positions
of the Sun he had to account correctly for the effect of solar parallax (the
difference of the Sun's apparent position at, say, noon and sunset after
accounting for its mean motion along the ecliptic). This depends on the
radius of the Earth and the distance to the Sun. He greatly underestimated
this distance, adopting a value of 1150 Earth radii.[32] Instead of the correct
value of 8″.8 for the solar parallax, Tycho used the Ptolemaic value of about

3'. This could lead to errors of a couple of arc minutes in his derived coordinates. (*c*) It must be kept in mind that Tycho's observations were all made with the naked eye. Fainter stars would be much more difficult to center properly in the sights. Though Tycho was the greatest naked eye astronomer of all time, though not the last,[33] and his results were the most significant at least since Ulugh Beg, his observations were also limited by the resolution of the naked eye (on the order of 1'). Even if he had much larger, more accurately aligned instruments, without the introduction of a telescope he had reached the limitation of the technology. (*d*) Tycho's favorite instrument, the mural quadrant, could not be used to observe objects north of the zenith. With the use of more than one instrument for the production of a catalogue of stars, there are extra problems in eliminating systematic errors in the combined results. Along these lines, in 1671 when Jean Picard, Erasmus Bartholin, and Ole Roemer went to Hven to measure the position of Tycho's observatory, they found that Tycho may have incurred an error of 18' in assigning the direction of the meridian.[34] (*e*) He states that the clocks he used 'are constructed in such a way, that they give not only the single minutes, but also the seconds, with the greatest possible accuracy, and imitating the uniform rotation of the heavens. Although it is difficult to make the clocks do this, one can, by exercising great care, to a certain degree attain this end'.[35] An error of only 4 s of time would translate to an error of 1' at the celestial equator, and we can safely regard any of his observations based directly on clock timings as not very accurate. Though Tycho's nine fundamental stars had positions known to better than ± 1',[36] for the above reasons we can regard his other positional measurements as no more accurate than ± 2', and probably no worse than ± 5'.

Also included in the *Progymnasmata* was Tycho's theory of the Moon's motion. Whereas Hipparchus had determined the first inequality of the Moon's motion in longitude (that it moved most rapidly at one part in its orbit and most slowly near the opposite part) and Ptolemy discovered the second inequality in longitude, the evection (to account for determinations of the eccentricity of the orbit at quadrature), Tycho discovered four inequalities in the Moon's motion, two in longitude and two in latitude. Concerning the motion in longitude, there was still a large error at the octants (when the Moon is 45° or 135° on either side of the Sun). The third inequality, amounting to about 40', is known as the *variation* and was discovered by Tycho during the years 1590–5.[37] During the winter of 1598–9, when he lived at Wandsbeck, near Hamburg, he discovered the fourth inequality, the so-called annual equation with a period of one solar year. Tycho's discoveries of the motion in latitude deal with the inclination of the Moon's orbit. He revised Ptolemy's value of the inclination from $5°$ to $5\frac{1}{4}°$, interpreting it as a long-term change rather than an error on the part of Ptolemy. He then noticed that his own observations had been made at quadrature, whereas most of the other observations had been concentrated at full moon or new moon. By 1595 he was able to show that the Moon's orbital inclination varied on a semi-monthly time scale.[38]

Tycho's last observation made at Hven is dated 15 March 1597. Since the death of his patron, Frederick II, in 1588, his fortunes had been on the decline. This was primarily his own fault. Even as a young man he had exhibited arrogance and held a very high opinion of himself. His celebrated duel with another Danish nobleman, one Manderup Parsbjerg, on 29 December 1566, is a case in point. The fight was presumably over who was the better mathematician. They duelled in the dark with sabres, and Tycho lost part of his nose, forever after wearing a prosthetic piece composed of a gold–silver–copper alloy. Tycho's nose is perhaps the most famous in science, being a principal object of study when his tomb in Prague was opened in 1901 by some Czech historians.[39]

Tycho had not treated his tenants at Hven well. Other Danish noblemen at court were jealous of Tycho's entertainment of visiting dignitaries. The great number of instruments built at Hven gives one the impression that instruments were built to create work (much like modern-day recipients of grants rushing to spend money before the end of the fiscal year in the fear that the grant may not be renewed in full). Some claimed that Tycho's poems, which he dedicated to good friends, were published to keep his paper mill busy.[40] This could hardly make a good impression on the new king, Christian IV (king 1588–1648, but crowned in 1596), who entertained a lower estimate of the value of astronomy than did Frederick II. Tycho was forced into exile, eventually settling in Prague under the patronage of Rudolf II, Holy Roman Emperor (1576–1612), the same patron for whom Kepler's *Rudolphine Tables* are named. Most of Tycho's instruments migrated with him, but they were never again properly set up. By 1600 the island of Hven had become an astronomical ghost town. Tycho's instruments survived until at least about 1620, but after that their fate is unknown.

Just as the observations of Hipparchus were used by Ptolemy to construct a careful theory of the heavens, so were the observations of Tycho used by Kepler. In January of 1600 the 29 year-old Kepler arrived in Prague. He too was an exile. Though their temperaments clashed greatly and Tycho jealously guarded his observations, by the time Tycho died on 24 October 1601 Kepler had learned enough about Tycho's data to realize he had the building blocks of a whole new celestial model. He concentrated on the motions of the planet Mars and soon discovered that, in its varying distance from the Sun, Mars swept out equal areas of its orbit in equal times. (This became known as his second law, though it was discovered first.) The best theoretical model he could produce still left errors of 8′, and Kepler knew that Tycho's data were more accurate than this. Instead of an eccentric (non-centered) circular orbit, Kepler began experimenting with non-circular orbits, using at first an egg-shaped curve ('*plani oviformis*').[41] By 1605 he discovered that the true form was indeed an ellipse, with the Sun at one focus. Kepler has carefully recorded his line of reasoning on this matter, producing 'the first instance in the history of science of a discovery being made as the result of a search for a theory, not merely to cover a given set of observations, but to interpret a group of refined measurements whose probable accuracy was a significant factor'.[42] Kepler's laws of planetary motion may be stated as follows:

1) The orbit of a planet is an ellipse with the Sun at one focus.
2) The radius vector of a planet orbiting the Sun sweeps out equal areas in equal times.
3) If we consider the periods (p) of the planets orbiting the Sun and their orbit sizes (a, technically the semi-major axes of the ellipses), the following relationship holds –

$$p^2 = \text{constant} \times a^3.$$

The first two laws were published in Kepler's *Astronomia nova de motibus stellae Martis* (Prague, 1609). The third law was published in the *Harmonice mundi* (Linz, 1619).

The consideration of Kepler's laws and Galileo's discourses on the laws of falling bodies led Newton to the notion of universal gravitation. Concerning the year 1666 Newton writes,

> I began to think of Gravity extending to ye orb of the Moon, & (having found out how to estimate the force with which a globe revolving within a sphere presses the surface of the sphere), from Kepler's rule ... I deduced that the forces which keep the Planets in their Orbs must [be] reciprocally as the squares of their distances from the centers about which they revolve.[43]

In 1684 Newton was able to show mathematically that an inverse square force would produce orbits that were conic sections (circle, ellipse, parabola, or hyperbola), of which Kepler's first law is a particular case. As laid out in Newton's *Philosophiae naturalis principia mathematica* (1687; 'the *Principia*'), the universe was once again subjected to elaborate mathematical modelling, but the difference between it and, say, the *Almagest* or even *De Revolutionibus*, was that the universe was now subjected to a foundation of physical and physically testable laws.[44]

Just as Ptolemy spawned generations of commentators, so did Newton, with such successors as J. L. Lagrange, who published his *Mécanique analytique* in 1788, and P. S. Laplace, who wrote the 5 volume *Mécanique céleste* (1799–1825). The discovery of the planet Neptune by Adams and Le Verrier (1846), the production of astronomical ephemerides, the refinement of planetary theory by Newcomb and others (see Chapters 4 and 5), putting a man on the Moon – all these have their basis in Newton's *Principia*.

What of observational astronomy after Tycho Brahe? Certainly the most significant development was the invention of the telescope, which was first applied to the study of the stars by Galileo (1609). Kepler, who made no telescopic observations of note due to bad eyesight, yet was in communication with Galileo, describes the telescope as a 'much-knowing perspicil, more precious than any scepter! He who holds thee in his right hand is a true king, a world ruler ...'.[45] If ever there were a good example of provocative discoveries to be made with the application of a new device, placing a telescope in the hands of Galileo during the winter of 1609–10 has to rank near the top. His *Siderius nuncius* (Starry messenger), published in Venice in March of 1610, revealed that the Moon was mountainous and the Milky Way was composed of many more stars than visible with the unaided eye, both contrary to Aristotelian principles, and that Jupiter had four moons, showing that all bodies do not have to revolve about the Earth. Galileo later discovered that Saturn's image had two 'knobs' (his telescope had insufficient resolution to resolve the rings). He saw that Venus exhibited phases like the Moon, as predicted by Copernicus. Galileo also observed sunspots.

However, Galileo, 'now often demoted to a status little above that of press agent for the scientific revolution'[46] (forgive me *Signor* Galilei!), had too many interests and became involved in too many controversies to establish regular astronomical observations on par with the activity of Tycho Brahe at Hven. The new tree planted by Galileo would bear the first mature fruit later in the seventeenth century at the observatories of Paris and Greenwich, to which we now turn.

4

Paris and Greenwich

There are tens of thousands of men who could be successful in all
the ordinary walks of life, hundreds who could wield empires,
thousands who could gain wealth, for one who could take up this
astronomical problem with any hope of success. The men who have
done it are therefore in intellect the select few of the human
race, – an aristocracy ranking above all others in the scale of being.
The astronomical ephemeris is the last practical outcome of their
productive genius.
Simon Newcomb (1903)

The observatories at Paris and Greenwich were both founded in the second
half of the seventeenth century. Still in existence today, their histories
constitute a great amount of activity in all of astronomy, only a small portion
of which can be touched on here. Their origins can both be traced to three
principal factors: (*a*) there was an increase of pure interest in science made
manifest at the end of the Renaissance and associated with the invention of
the telescope; (*b*) the development of navigation as an element of national
importance demonstrated the necessity of state-funded astronomical institutions; and (*c*) the increased prosperity due to trade (via navigation) led to a
rivalry among various countries regarding sometimes ostentatious de-
monstrations of the splendor of their Courts and a rivalry among them for the
promotion of the arts and sciences.

Following the lead of the *Accademia dei Lincei*, founded in Rome in 1603
and counting among its early members Galileo, the first telescopic observer of
the skies, the French and English soon founded scientific circles of their own.
In France Father Marin Mersenne organized meetings at his home beginning
in the late 1630s. The circle included such famous intellects as Descartes,
Gassendi, Etienne Pascal and his son Blaise Pascal. In London the Royal
Society was established by Royal Charter in 1662, counting among its
members Christopher Wren (architect), Robert Boyle (chemist), and later
(1672) Isaac Newton, who served as its president from 1703 to 1727. The
Royal Society's first patron was King Charles II. Meanwhile, in France the
scientific luminaries decided to obtain official status for their group. This led
to the founding of the *Académie des Sciences* in 1666,which embodied the
splendor of the reign of Louis XIV (Fig. 4.1).[1,2]

Fig. 4.1. King Louis XIV visits the Paris Observatory, 1682. From Wolf, C., *Histoire de l'Observatoire de Paris de sa Fondation à 1793*, Paris: Gauthier-Villars, 1902 (frontispiece.)

Fig. 4.2. View of the Paris Observatory. From Wolf (1902), Plate X.

Fig. 4.3. Another view of Paris Observatory. From Wolf (1902) Plate II.

Of the French and the English, the former were the first to plan the construction of an observatory (Figs 4.2 & 4.3), which was designed by Claude Perrault, one of the architects of the Louvre. It consisted of a flat 27 m terrace and two octagonal towers. Observational equipment was to be set up on the roof, or observations could be carried out through the windows of the building. For a window facing south, this would allow transits of objects such as planets to be observed in order to obtain accurate celestial coordinates. However, a moment's consideration allows one to perceive how inadequate this arrangement would be for actual astronomical observations. Instruments set up on the roof were not sheltered from the wind, and they could not be mounted there permanently (due to rain and snow), something necessary for the accurate alignment and collimation of an instrument. Observing through a window precludes observing throughout the whole plane of the meridian, also necessary for proper collimation of the instrument. Similarly, that would exclude observing at the zenith, where accuracy is more easily attainable due to the negligible effect of atmospheric refraction.

Construction of the observatory was begun in 1667 and though some aspects dragged on until 1682,[3] the edifice was basically finished by 1672. The first director (though this was an unofficial position until 1771) was Gian Domenico (Jean Dominique) Cassini (Fig. 4.4), the first of four generations of Cassinis to fill that position. Cassini arrived from Bologna in 1669 when the observatory was built only as far as the second floor. He suggested a couple of improvements in the construction. Instead of a third octagonal tower on the northern side of the observatory, a square tower was built to house a quadrant.[4]

> But the modifications proposed by him, shewed that his ideas were not in accordance with the existing requirements of practical astronomy. Among the alterations recommended by him, was the construction of an apartment in which the celestial bodies might be observed from east to west throughout the whole of their diurnal courses. Such a mode of observation might have been very serviceable for the discovery of satellites, and for the examination of the physical constitution of the planets, but in so far as the determinations of positions of celestial bodies was concerned the tendency of the age was towards observations made exclusively in the plane of the meridian with fixed instruments.[5]

What Cassini wanted was in essence a modern observatory dome, an extension of the idea partly put into practice for some of Tycho Brahe's instruments (Chapter 3). Though Tycho had also improved the state of meridian instruments, it was developments in the late seventeenth century that led to the age of positional astronomy. Planetary astronomy then and

Fig. 4.4. Jean Dominique Cassini (1625–1712), first director of the Paris Observatory, Photo courtesy L'Agence Roger-Viollet, Paris; Original in the Bibliothèque Nationale.

now goes though periods of popularity, and with the rapid improvement of telescopes at that time, why not observe everything in the sky? Cassini is best known for his planetary observations, which included accurate determinations of the rotational periods of Mars and Jupiter, the discovery of the gap in Saturn's rings known as Cassini's Division, and the discoveries of four of Saturn's satellites (Iapetus, Rhea, Tethys, and Dione). His improvements in the tables of Jupiter's satellites enabled Ole Roemer to discover that the velocity of light was finite (see below).

Because of the inadequacies of the building and the pomp of the French Court, the Paris Observatory has been called 'l'Observatoire de parade'. 'Built, it must be said, at a time when observational astronomy was in a transition state, and ideas were not as well-defined on what the science demanded, as they were but fifty years later, the Paris Observatory has steadily defied adaptation of the original structure to purposes of observation'.[6] Thus wrote an American astronomer at the end of the nineteenth century.

Nevertheless, great scientific progress was achieved. One of the most significant results arose inadvertently from a prime practical endeavor – the accurate determination of longitude. The determination of latitude is fundamentally an easy matter, for it is the elevation of the north celestial pole above the horizon. The determination of longitude is a time difference between two places on the Earth. As the Earth turns once in 24 hours, 1 hour corresponds to 15° of longitude, 4 minutes of time correspond of 1° of longitude, 1 minute of time corresponds to 15′ of longitude, etc. In order to determine longitude one in principle compares the local time (by the stars or by the Sun) with the time at a standard meridian (say Paris or Greenwich). The difference of local standard time or local sidereal time is directly related to the difference of longitude. For navigation at that time, no clocks existed which could keep accurate time at sea (in order to bring along the standard time), and the determination of local sidereal time is best done with accurate star catalogues and accurately mounted telescopes. It occurred to Galileo that timings of eclipses of Jupiter's moons by Jupiter could be used for fixing absolute moments in time. Others soon suggested eclipses of our Moon for such time references.

Jean Baptiste Colbert, principal financial advisor to Louis XIV and a prime mover behind the establishment of the French Academy of Sciences and Paris Observatory (seen in Fig. 4.1), assigned to the Paris astronomers the task of improving the tables of ephemerides of Jupiter's satellites. J. D. Cassini, who had published his set of tables for the Jovian satellites in 1668 on the basis of his observations in Italy, and his nephew Giacomo Filippo Maraldi were the first to note that the times of the eclipses of the satellites deviated from the predictions. (This was later attributed to the ellipticity of Jupiter's orbit.) Ole Roemer, a Danish astronomer invited to Paris, used these tables of eclipses to make further observations and discovered that the inequality seemed to depend on the distance between the Earth and Jupiter. Though Galileo had been convinced of the finiteness of the speed of light, Cassini rejected that notion. Both of their opinions were *a priori* ones, however, and it was Roemer's extensive treatment of the eclipses, including how future events would deviate from the tables, that correctly demonstrated Galileo's notion. However, for Roemer it was enough to demonstrate the finiteness of the speed of light. He did not calculate a value for the speed, though he did state that it took light 22 minutes to cross the diameter of the Earth's orbit.

The common belief is that 'Roemer's value of the speed of light' was not accurate because of uncertainties in the size of the Earth's orbit, but these calculations were performed by others.[7] Many scientists were not convinced of Roemer's interpretation because of the weight of Cassini's opinion and that of various Cartesians, but after Bradley's discovery of aberration (see p. 86), the notion of the finiteness of the speed of light has prevailed.[8]

One of the most unusual features of the early years of Paris Observatory was the use of long-focus refractors. Because the exact focal length of a lens depends on the wavelength of light (blue light is refracted more than red light), single-lens refractors suffer from serious chromatic aberration. The early refractors also had lenses which had spherical surfaces. This led to spherical aberration – even if all the rays falling on the lens were of the same color, they still would not all come to a single focus. Thus the early telescopic images were marred by 'beautiful' colored halos and the details were somewhat smeared out. The opticians of that day discovered, however, that these two detrimental effects were reduced if the focal length of the objective were increased relative to its diameter. Whereas a modern amateur telescope might have a 6-inch diameter objective of 48 inches focal length (f/8), the opticians of the mid-seventeenth century produced lenses with focal ratios on the order of f/150.[9]

Some of the principal users of long-focus telescopes were Johannes Hevelius of Danzig (who mapped the Moon), Christiaan Huygens (who was one of the original members of the French Academy of Sciences), and Jean Dominique Cassini. The Italian lens makers Eustachio Divini and Giuseppe Campani produced the finest long-focus lenses made at that time. Once at Paris, Cassini used object glasses with focal lengths of 17, 34, 37, 80, 90, 100 and 136 feet by Campani, a 4-inch f/111 lens (focal length = 37 feet) by Divini, and lenses of 40 and 77 feet by Giovanni Borelli.[10] Cassini mounted the shorter focal length lenses in a tube supported from a mast on the observatory terrace. For the larger focal length lenses he made use of a water tower moved to the observatory for exactly this purpose (Fig. 4.5).

Lest the reader think that the giant telescopes were a peculiar manifestation of the late seventeenth century only, consider a modern reflecting telescope with a coudé (elbow) flat. This optical arrangement is used for providing a long effective focal length for high-resolution spectroscopy by passing the light down the polar axis of the telescope to equipment situated in another area of the building. The average coudé spectrograph is also too heavy to mount directly on the telescope itself. This type of arrangement was pioneered at the Paris Observatory in the 1880s under the direction of Admiral Mouchez. Thus, there is a remnant of the previously unwieldy

Fig. 4.5. The Marly tower, moved to the Paris Observatory and used to support
ultralong-focal length refractors. From Wolf (1902), Plate XI.

behemoths used by Cassini in the most modern telescopes used today.

The astronomers at Paris Observatory were pioneers in the area of geodesy – the accurate determination of geographical positions on the Earth and the determination of the shape of the Earth. The application by Picard and Roemer of telescopes to measuring instruments such as transit instruments, and Picard's use of the pendulum clocks recently invented by Huygens allowed stellar positions, time determinations, and geographical coordinates (all of which are interrelated) to be measured with a great increase of accuracy. Whereas Tycho's instruments allowed measurements to be made to an accuracy of a few minutes of arc, with the new French instrumentation the accuracy achievable for differential measurements was on the order of a few seconds of arc.

Jean Picard, one of the founders of the Academy of Sciences in 1666, worked at the Paris Observatory from 1673 until his death in 1682. In 1668 he was placed in charge of the first measurement of an arc of the meridian, from Paris to Amiens, amounting to 1° 21′. The results were published in 1671 in his book *Mesure de la Terre*. In 1683 Cassini and Philippe de la Hire began an extension of this meridian measurement to include all of France (8° 30′). This project was not completed until after 1700 when the arc measured by Cassini and Maraldi from Paris to Collioure indicated that the Earth had a diameter greater at the poles than at the equator. This idea was supported by the Cartesian philosophers and was contrary to Newton's conclusion that the Earth should be an oblate spheroid (smaller in diameter at the pole), much like Jupiter appears in a small telescope.

The French Academy later sent expeditions to Peru (1735) and Lappland (1736), the results of which clearly showed that the Earth was oblate. To complicate matters, however, Nicolas de Lacaille went to South Africa in 1751, where he measured the positions of southern stars and also carried out a geodetic survey, measuring a meridian arc of 1° 12′. His survey indicated that the Earth, while flattened at the poles, was pear-shaped, with the southern hemisphere larger.* In carrying out this type of work one uses a plumb line to determine the level position, and Lacaille failed to take into account the effect of nearby mountains on the plumb line. This problem had been considered at that time by writers such as Pierre Bouguer in his *La Figure de la Terre* (1749). The re-execution of Lacaille's survey, begun by Thomas Maclear in 1837, indicated that the total baseline was actually shorter than previously measured (decreasing the anomaly by half), and after increasing the arc of the meridian to 3° 34′, which eliminated the effect of the previous plumb line perturbations, Maclear found a length of a degree almost exactly that derivable from meridian observations in the northern hemisphere.[11]

Wolf's detailed *Histoire de l'Observatoire de Paris de son Fondation à 1793* covers the history of the observatory up to the period of the French Revolution. During the first 120 years of its existence Paris Observatory was directed by four succeeding generations of Cassinis, and the work there primarily involved positional astronomy though, because of the state of the observatory and the instruments and the way the work was organized, they did not carry out the type of multi-year projects undertaken at Greenwich Observatory where there was a clearly defined observational program, a director, his assistants and calculators, all supervised by a committee of academicians. Whereas Greenwich Observatory was founded specifically for the goal of determining longitude, the Greenwich Board of Visitors was created in 1710, and the Board of Longitude in 1714, at Paris it was not until 1795 that their *Bureau des Longitudes* was organized to improve navigation and supervise the work at Paris Observatory. By this time, however, due to the development of the marine chronometer by John Harrison, the problem of longitude at sea had largely been solved. Still, the French were far ahead (timewise) when it came to the publication of ephemerides – tables of

---

* Modern satellite measurements indicate that the Earth is pear-shaped, but Lacaille's measurements could not show the amount of 'pear-shapedness' now accepted (I thank Alan Batten for pointing this out).

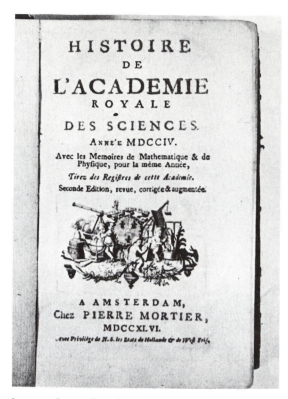

Fig. 4.6. Title page of an early volume of the *Memoires* of the Royal Academy of Sciences, Paris.

positions of the Sun, Moon and planets. The first volume of the *Connaissance des Temps* was for 1679. By contrast, the British *Nautical Almanac* was first published in December 1766, and the *American Ephemeris* first appeared in 1855. Similarly, the annual volume of the activities of the *Academie des Sciences*, such as that shown in Fig. 4.6, was one of the first serial scientific publications, in essence a scientific journal.* Naturally, descriptions of astronomical observations carried out at the Paris Observatory are to be found there.[12]

The work at Paris Observatory during most of the nineteenth century involved celestial mechanics. The most famous achievement was the prediction by U. J. J. Le Verrier in 1846 of the location of a possible new planet outside the orbit of Uranus. The problem arose because it appeared that Uranus' motion was being perturbed by some unknown body. The position of the new planet (Neptune) was also independently calculated by

---

* The first two scientific journals were the French *Journal des Sçavans* and the British *Philosophical Transactions of the Royal Society of London*, both of which first appeared in 1665.

Fig. 4.7. Astrographic refractor made by Paul and Prosper Henry, Paris. Refractors based on this telescope were used for the international *Carte du Ciel* project. From Winterhalter, Albert G., *The International Astrophotographic Congress and a Visit to Certain European Observatories and Other Institutions*, Appendix I to Washington observations for 1885, Washington, DC: Government Printing Office, 1889, facing p. 103.

John Couch Adams in England. However, Adams' work was unpublished and his results were only known to a select few in England, where the search for the new planet was not carried on with much vigor. Meanwhile, on the basis of Le Verrier's calculations, the new planet was optically revealed on 23 September 1846 by Johann Gottfried Galle at the Berlin Observatory. This observational success was due in large part to the existence in Berlin of some not yet published star charts. It was a simple matter to compare the actual field to see if an interloper had entered the area or if one of the stars on the chart had moved. Because Le Verrier had published his method of calculation in a series of three papers, while Adams' work was only known informally in England, many French felt that Le Verrier was being denied proper credit for his work after the British claimed that Adams' work had actually preceded that of Le Verrier. History has recognized that both men are to be considered the discoverers of Neptune.[13]

When Admiral Mouchez became director of Paris Observatory in 1878 following the death of Le Verrier, it was clear that observational astronomy would gain prominence over celestial mechanics. The late nineteenth century saw the development of astrophysical methods, in particular, photography. Due to the success of the brothers Paul and Prosper Henry in photographing faint stars and also the work of David Gill in South Africa, in 1887 the Paris Observatory hosted an international astrophotography congress in order to organize a scheme for photographing the whole sky. This would provide a solid record of the star positions and magnitudes which could later be supplemented by additional photographs for studies of stellar motions.

It was resolved to produce two sets of plates. The longer exposure photographs (about 15 minutes) would reach 14th magnitude. The shorter exposures (2 minutes) would record stars brighter than 11th magnitude for the compilation of a catalogue of stars. The instruments to be used would be like the photographic refractor of the brothers Henry (Fig. 4.7), which had an objective size of 33 cm (13 inches) and a focal length of 3.43 m (giving a focal ratio of f/10.4). The focal length was chosen to give a plate scale of 1 minute of arc per millimeter. It was decided to have the plates cover 2° on a side but overlap such that a star at the corner of one plate would be in the middle of another plate. There would be approximately 4 million stars for the catalogue and 25 million stars on the longer exposure plates. To photograph the sky twice over with plates of four square degrees takes over 20 000 plates. Because of worries about shrinkage of the emulsion on the plates, prior to photographing a patch of sky, each plate was to be exposed to an artificial grid. To eliminate spurious images, each plate was to be exposed

Table 4.1. *Participants in the* Carte du Ciel *(sky map) project*

Zone	Original Observatory (1891)	Replacement
+90° to +65°	Greenwich	
+64° to +55°	Rome-Vatican	
+54° to +47°	Catania (Italy)	
+46° to +40°	Helsingfors (Helsinki)	
+39° to +32°	Potsdam	+39° to +36° Hyderabad (India)
		+35°, +34° Uccle (Belgium); Paris
		+33°, +32° Hamburg; Paris
+31° to +25°	Oxford	
+24° to +18°	Paris	
+17° to +11°	Bordeaux	
+10° to +5°	Toulouse	
+4° to −2°	Algiers	
−3° to −9°	San Fernando (Spain)	
−10° to −16°	Tacubaya (Mexico)	
−17° to −23°	Santiago (Chile)	Hyderabad
−24° to −31°	La Plata (Argentina)	Cordoba (Argentina)
−32° to −40°	Rio de Janeiro	Perth
−41° to −51°	Cape of Good Hope	
−52° to −64°	Sydney	
−65° to −90°	Melbourne	Sydney

three times with slight offsets; thus the stellar images are all triangles of three images.

Though the *Carte du Ciel* project was carried out by many observatories (Table 4.1), and it was felt at the start that it would take perhaps 10 years to complete; many parts were not completed and the project officially ended in 1970. The *Astrographic Catalogue*, which contains the positional inform-ation, usually only gives *x*- and *y*- coordinates for the stars. Only for the declinations +40° to +46° (Helsingfors) and +54° (Catania) has the transformation to right ascension and declination been completed.[14]

One may sensibly conclude that the *Carte du Ciel* was prematurely planned. Though photography had made great strides in the late nineteenth century, the participants of the Paris astrophotography congress might have predicted that further progress was possible in optics and photographic emulsions. For example, consider the National Geographic Society-Palomar Observatory Sky Survey. It covers the sky from declination +90° to −33°; the limiting magnitude is 21.1 for the plates to declination −21° and magnitude 20 for those to $-33°$; and it was carried out by a single telescope in a period of 10 years (1949 to 1958).

In addition to the astrophotography conference of 1887, the Paris Observatory was also host to a conference on astronomical constants in 1896, at which values for the distance to the Sun and reduction parameters such as nutation and aberration (see below) were agreed upon. The Paris Observatory director Benjamin Baillaud played an important role in the founding of the International Astronomical Union, serving as its first president from 1919 to 1922. Undoubtedly, the importance of the Paris Observatory must have figured prominently in the decision to make French one of the two principal languages of the IAU, along with the fact that French was still the principal diplomatic language in 1919.

In 1926, the Paris Observatory was united with the Meudon Observatory, which had been founded in 1876 by Jules Janssen and which was one of the first centers for astrophysics. Thus, twentieth century astronomy in France embodies the dual traditions of positional astronomy and astrophysics. Today the staff of the Paris Observatory numbers more than 700 scientists, technicians, and administrators working at Paris, Meudon and elsewhere on all aspects of astronomy.

The volume of historical material pertaining to the Royal Greenwich Observatory (RGO), both published and unpublished, is immense. The most complete single source of information is the three-volume history of the observatory published in 1975 to commemorate the tercentenary of its founding.[15] That is just a beginning of what could be said about Greenwich Observatory, however. For example, the seventh Astronomer Royal, G. B. Airy, was so preoccupied with methodical record-keeping that it was said if he wiped his pen on a piece of blotting paper he would sign, date, and file it for possible future use.[16] Consequently, the Airy Collection at RGO amounts to over 700 volumes of notes and correspondence.[17] Though Airy did serve as Astronomer Royal for many years (see Table 4.2), one can imagine how much material there is altogether for the entire history of the observatory. Yet, curiously, from its founding until this century the activity at Greenwich Observatory was not the result of a large number of people. From 1675 to 1900 the *total* number of established staff amounted to less than 50.[18] The number of staff in 1983 amounted to 237, but they have recently been required to reduce the staff size.

As mentioned earlier in this chapter, the determination of longitude is particularly important for navigation. Tables of eclipses of the moons of Jupiter were compiled for this purpose, but that method had its problems in that the observed times of the events in question depended upon the

Table 4.2. *Directors of Greenwich Observatory*[a]

John Flamsteed (1646–1719)	1675–1719
Edmond Halley (1656–1742)	1720–1742
James Bradley (1693–1762)	1742–1762
Nathaniel Bliss (1700–64)	1762–1764
Nevil Maskelyne (1732–1811)	1765–1811
John Pond (1767–1836)	1811–1835
George Biddell Airy (1801–92)	1835–1881
William Henry Mahoney Christie (1845–1922)	1881–1910
Frank Watson Dyson (1868–1939)	1910–1933
Harold Spencer Jones (1890–1960)	1933–1955
Richard van der Riet Woolley (1906–1986)	1956–1971
E. Margaret Burbidge (1922–)	1972–1973
Alan Hunter (1912–)	1973–1975
F. Graham Smith (1923–)	1976–1981
Alexander Boksenberg (1936–)	1981–

[a] Until 1971 the positions of Astronomer Royal and Director of Greenwich Observatory were one and the same. From 1972 to 1982 the Astronomer Royal was Martin Ryle. Since 1982 the Astronomer Royal has been F. Graham Smith.

magnification and resolution of the telescope,[19] and for a couple months of the year Jupiter is on the other side of the Sun from the Earth and is lost in the Sun's glare. In the early sixteenth century Johann Werner, among others, suggested that the 'method of lunar distances' be used for the derivation of longitude. In this method the position of the Moon is determined with respect to one or more (typically two or more) bright stars. From a comparison of the local time at which the Moon achieves a particular position with the predicted time referenced to a known longitude, the longitude at which the observations were made could be derived.[20] However, this assumes that one has accurate positions for the stars and adequate tables of the Moon's position, neither of which were to be had before the founding of Greenwich Observatory.

In 1674 a Frenchman who called himself the Sieur de St Pierre announced to the English Court of Charles II that he had a method of obtaining the coordinates of the Moon by the method of lunar distances. The following information had to be provided: (*a*) measurements of the heights above the horizon of two stars, in degrees and minutes; the closer these stars were to the celestial equator, the better; (*b*) the elevation of the pole in degrees and minutes (i.e. the latitude); (*c*) the height of the upper and lower limbs of the Moon in degrees and minutes; (*d*) it must be stated whether the two stars are in the eastern or western half of the sky at the time of observation.

Fig. 4.8. John Flamsteed (1646–1719), the first Astronomer Royal. Photo courtesy of Christopher St J.H. Daniel.

The elevation angles were all to be given for the same instant in time, something which can be obtained without much difficulty by a few observations of each and some interpolation. The problem of the refraction in the Earth's atmosphere was not addressed, nor that of the effects of the diurnal parallax of the Moon. A Royal Warrant was issued in December of 1674 to a committee headed by Lord Brouncker, the Bishop of Salisbury. Among the members of the committee were Robert Hooke, Sir Christopher Wren, and a young assistant named John Flamsteed (Fig. 4.8), who had begun his own astronomical observations at Derby in 1662 at the age of 16 and who had achieved the reputation as a skilful observer. The committee met on 12 February 1675. Five days later Flamsteed handed over some observations in his possession, dated 23 February 1672 and 12 November 1673, which would suffice for a test of the Sieur's method. Instead of such a test, the Frenchman complained that these were calculations, not observations (i.e. fake data). This further convinced Flamsteed that the Sieur's intentions at Court were for political gain and not based on scientific interests. The Sieur's lack of understanding of the basics of positional astronomy implied that 'his' method had been borrowed from other persons who had investigated the matter, such as Jean Baptiste Morin, who had tried to interest the French Court in it in 1634, and who had been told that the method was not yet practical.

At that time the best tables of the Moon's motion could be in error by 12′. The positions of the stars in Tycho Brahe's catalogue were to be used and assumed to be absolutely correct, but Flamsteed had found errors in Tycho's positions based on his own observations that amounted to 5′ or more.[21] The best that could be done on the basis of the method of lunar distances, given the instruments and tables of that day, was an accuracy of no better than 3° in longitude (about ± 110 nautical miles at the latitude of London). What was needed was a systematic recataloguing of the stars and also better tables of the positions of the Moon and planets.

By a Royal Warrant of 4 March 1675 Flamsteed was appointed Astronomer Royal, at a salary of £100 per year. His task was 'forthwith to apply himself with the most exact care and diligence to the rectifying the tables of the motions of the heavens, and the places of the fixed stars, so as to find out the much-desired longitude of places for the perfecting the art of navigation'.[22]

Another warrant of 22 June 1675 called for the establishment of 'a small observatory within our park at Greenwich, upon the highest ground, at or near the place where the castle stood, with lodging-rooms for our astronomical observator and assistant...'[23] The observatory was to be designed by Christopher Wren, who had recommended Greenwich as the site. The cornerstone of what became known as Flamsteed House (Fig. 4.9), built 'for the Observator's Habitation, and a little for Pompe' (Wren's words),[24] was laid on 10 August 1675, and the building was finished in less than a year. Flamsteed took up his residence there on 10 July 1676 and soon began his observational program.

Before we proceed with further discussion of the history of Greenwich Observatory, let us briefly describe some types of instruments that were to be used there. A *transit instrument* is a telescope mounted on a horizontal (east–west) axis and able to observe objects on the meridian – the great circle on the sky passing from the south point on the horizon to the zenith, through the north celestial pole and to the north point on the horizon. (Naturally, in the southern hemisphere the south celestial pole is above the horizon and on the observer's meridian.) The eyepiece end of a transit instrument is usually provided with a set of parallel wires running up and down in the field of view and one horizontal wire. One aligns the instrument such that a star passes

Fig. 4.9. Flamsteed House at the Old Royal Observatory (early twentieth century). Photograph courtesy Royal Greenwich Observatory, Herstmonceux.

along the horizontal wire from east to west and times the passage of the star across the parallel wires. Given the separation of the wires, it is possible to derive a more accurate value of the time of transit across the central wire than would be obtained with just a single measurement. If the instrument is set up with a graduated elevation circle, it is called a *transit circle*. One would then be able to measure the zenith angles of the stars and derive their declinations in addition to the right ascensions obtainable from the transit times. A *mural arc*, or *mural circle*, is a graduated arc of a circle mounted on a wall and containing a not very powerful telescope for sighting a star. If aligned in the north–south direction, it is also called a *meridian arc*. With this type of instrument it is possible to derive right ascensions and declinations, but meridian arcs were used more often for determining declinations. Transit instruments, which usually had more powerful telescopes, were customarily used for obtaining right ascensions. It was a matter of being able to align an instrument for a principal task in order to obtain the most accurate results possible.

A *zenith sector* is like a meridian arc, but can only be used to observe objects near the zenith. The focal length of a zenith sector is usually much longer than a telescope for a mural arc or transit instrument. The resultant plate scale (the number of millimeters corresponding to, say, 1 arc minute at the focus) is therefore much greater, allowing the measurement of positions with greater resolution and accuracy. Use of a zenith sector eliminates the necessity of correcting the observations for the effects of atmospheric refraction. Because the light passing at an oblique angle through the atmosphere must pass through more and more air as the line of sight gets closer to the horizon, and because the density of the atmosphere decreases steadily with altitude, the curved path that a light ray takes as it enters the atmosphere is more and more curved as it is closer to the horizon. Refraction makes all objects look higher in the sky. At an elevation angle above the horizon of 45° the *refraction correction* is about 1′, while at the horizon it is about 35′. Because the refraction is dependent upon the density of the air, and this depends on the temperature and pressure (i.e. the weather), it is not possible to provide refraction tables which are always valid, even at a single site. A telescope on a high mountain top would be above more of the atmosphere, and the observations would have to be corrected for different, though less severe, effects of refraction.

Returning to the early history of Greenwich Observatory,[25] it was Sir Jonas Moore, the Surveyor General of the Board of Ordnance, who was responsible for making sure that Flamsteed received his salary. For some time Moore had served as Flamsteed's patron, without which little actual work would have transpired at Greenwich. Though Flamsteed was appointed Astronomer Royal, there was no provision for the purchase of instruments. As early as 1671 Moore had provided Flamsteed with a micrometer for determinations of the position of Mars in an investigation of the scale of the solar system. In 1675 Moore consented to have a mural quadrant built by Hooke for the new Greenwich Observatory. This proved to be inadequate, and a 7 foot iron sextant was then made by Edmund Silvester and Thomas Tompion. This, along with two pendulum clocks built by Tompion (also funded by Moore) and a 3 foot radius wooden quadrant and two small telescopes brought by Flamsteed from Derby, was the principal equipment used in the first years at Greenwich. The iron sextant was similar to the quadrant by Hooke in that the outer edge was cut like a gear and the arc was turned by means of a wormwheel. To measure angles one counted the wormwheel turns. The iron sextant was an equatorial instrument (i.e. it could be used to observe objects anywhere in the sky, not just on the meridian.) It had telescopic sights and was used to measure the angular separation between objects. It required two observers. With it Flamsteed had intended to triangulate all across the sky and measure the positions of the stars in the manner of Tycho Brahe, but he

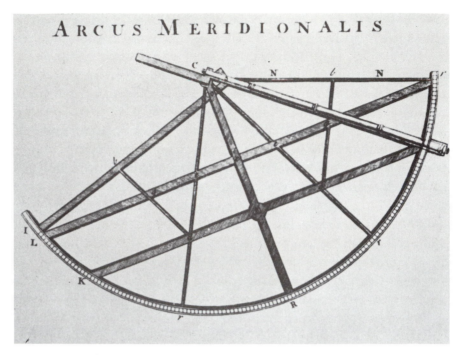

Fig. 4.10. Flamsteed's 7-foot mural arc. Photo courtesy of the Royal Greenwich Observatory, Herstmonceux; original from Flamsteed's *Historia Coelestis Britannica*, 1725.

soon found that the teeth of the arc were being worn, leading to unacceptable errors of up to 1'. Flamsteed then engraved a conventional set of divisions on the brass limb of the instrument so that all observations could be double-checked.[26]

The first observing assistant at Greenwich was Thomas Smith, who worked there until 1694. He was the 'other observer' for most of the observations on the iron sextant. His salary was a mere £26 per year, but this was eventually supplemented by an additional £34 per year paid out of Flamsteed's pocket. The observer's life was certainly not glamorous, and many observers were regarded by the Astronomers Royal as mere drudges.[27]

The iron sextant was used until 1689 when a 7 foot radius mural arc constructed by Abraham Sharp (Fig. 4.10) was finished. It cost Flamsteed £120 of his own money. Especially after the death of Moore in 1679, Flamsteed's financial situation was quite strained; he had been forced to take on students for a fee in order to make ends meet. Only after Flamsteed's father died in 1688, leaving him an inheritance, could Flamsteed afford this most important instrument which he had desired since 1676. He used it for the measurement of zenith distances (yielding declinations) and also timed meridian transits (yielding relative right ascensions). From a study of the Sun's position near the time of the equinoxes (September and March) he was able to determine the equinox points. As right ascension is the amount of time an object lies east of the vernal equinox, determining the absolute right ascension of one star yielded the absolute right ascensions of all the others. The meridian arc was mounted on the west wall of the quadrant house. Because it was fixed on the meridian, errors of collimation could regularly be checked, and it was possible to obtain more accurate star positions with it than with the equatorial sextant. From 1689 to 1719 a total of 28 650 observations were made with it.

About 1700 Flamsteed made a summary of the work he had carried out in his first 25 years as Astronomer Royal.[28] He had computed tables of the Sun's position useable for many years. He had observed about 3000 stars and had reduced the observations of half of them. From observations with the iron sextant of 1678 to 1681 he had provided Newton with the data to derive refraction tables. In 1694–5 Newton had also been provided with observations of the Moon with which he had hammered out a theory of the Moon's motion based on his recently (1687) published *Principia*. Flamsteed had also made observations of the Jovian satellites for possible use in longitude determinations. There remained the data reduction of half the stars observed, 400–500 lunar observations made with the iron sextant, and 300 observations of the Moon with the meridian arc. Then Flamsteed's principal task, the accurate cataloguing of the stars and observations of the Moon accurate enough to help determine longitude, would be finished. This would seem a simple task today: reduce a couple thousand observations and publish them. But what transpired was a mess, and it would be another 25 years till the publication of Flamsteed's data was complete.

Though Flamsteed had sent Newton many observations of the Moon, the latter appeared unsatisfied with the arrangement. Newton, as Airy would later, considered the taking of observations a task anyone could do. Newton felt that he had greatly helped Flamsteed regarding the reduction of the observations, not only of the Moon, but of the planets and for the production of his star catalogue; further, that Flamsteed owed him more observations and more respect. In 1695 Newton wrote Flamsteed a rude letter which said this, mentioning also that he had made comments along these lines to others. Flamsteed replied, 'I have just cause to complain of the style and expression in your last letter.'[29] The Greenwich observations had been carried out with instruments that Flamsteed considered his own personal property. As a result he felt that the data were his, with which to do as he pleased. The task at hand was too important to rush.

In 1703 Newton became President of the Royal Society. The following year at a metting of the Society it was unanimously decided to press Flamsteed for the publication of his observations. The cost of publication was to be born by the Society. An edition of 400 copies was to be printed. This work commenced in May of 1706, but the task was not completed until 1712. In the meantime, in December of 1710, the Board of Visitors was called into being, which required the Astronomer Royal to make available the observations of each year within 6 months after the year's end. One can understand the desire of the Royal Society to have Flamsteed's observations published, but it would have helped Flamsteed more if he had been provided with financial assistance for assistants and calculators. The appointment of this committee only added to the pressure Flamsteed felt.

The publication of the Royal Society edition of Flamsteed's data was considered so unsatisfactory by him that in 1715 he obtained the unsold copies of the two-volume work (300 of the run of 400) and promptly burned them. He resolved to publish his life's work at his own expense, an undertaking he did not live long enough to complete. He died in 1719.

The three-volume *Historia Coelestis Britannica* was published in 1725. As the name implies, it contains more than just Flamsteed's observations. The first volume contains observations of William Gascoigne and William Crabtree from 1638 to 1643 and Flamsteed's observations at Derby and Greenwich through 1689. The second volume contains the mural arc observations from 1689 to 1719, including lunar and planetary positions. The third volume contains a summary of the progress of astronomy and a description of Flamsteed's instruments and procedures. Then follows the star catalogues from all astronomers before Flamsteed and his own catalogue of 2935 stars, giving the right ascensions and polar distances (i.e. 90° minus the declinations), as well as the celestial latitudes and longitudes (ecliptic coordinates) of the stars, reduced to an epoch of 1689.[30] This star catalogue was the greatest achievement since the work of Tycho, and given the constraints under which Flamsteed worked, it must be regarded as one of the principal accomplishments in the history of astronomy which was primarily the work of one person.

Because Flamsteed considered the instruments at Greenwich his personal property (with just cause), upon his death Flamsteed's widow removed them from the observatory. They have not been seen since. Thus, when the 64 year-old Edmond Halley (of Halley's Comet fame) was appointed the second Astronomer Royal in 1720, he found himself in much the same position as his predecessor had in 1675 – he was director of an observatory without the means to observe. Unlike Flamsteed, however, Halley was granted £500 for the purchase of instruments. (This, although better than nothing, was still one-fourth the sum Flamsteed had spent as Astronomer Royal on instruments and assistants.) In 1721 Halley installed a transit instrument reportedly built by Hooke. The instrument, a small building to house it, and a clock to be used with it cost £61. The transit instrument was $5\frac{1}{2}$ feet long and was used for timing objects as they transited the meridian in order to determine relative right ascensions. This was Halley's principal instrument until 1725 when an 8 foot iron quadrant built by George Graham was installed in a new quadrant room. Halley had intended to procure two quadrants, one facing south and one facing north, but the cost of the building and the first quadrant amounted to over £300 and his grant could not cover the purchase of the second quadrant. Thus he was limited to observing the south half of his meridian. The single quadrant served him well, however, and was used as the prototype for many quadrants built by John Bird and Jeremiah Sisson during the years 1750–90, which were used all over Europe. The quadrant was more accurate than previous instruments because it was sufficiently rigid and was outfitted with two sets of divisions (like Flamsteed's iron sextant) to provide a double check on any measurement.[31]

Halley used the quadrant for measuring transits of objects on the meridian and also for measuring zenith distances. His principal object of interest was the Moon, which he assiduously observed for more than 18 years. Just prior to Halley's appointment as Astronomer Royal he had intended to publish tables of the Moon's predicted position but, upon becoming Astronomer Royal, he retained these for his own use. His program of observations of the Moon was intended 'to make the method of finding the longitude at sea by the moon more practicable than it is at present'.[32] A comparison of the observed and predicted places of the Moon from his data of 1722 to 1739 exhibits errors in celestial longitude frequently less than 1', but in some cases 5'–6'.[33] The pot of gold at the end of the rainbow in this case was the £20 000 first prize for being able to determine geographical longitude at sea to within $\frac{1}{2}$° or 30 nautical miles.

The Board of Longitude had been created in 1714 to judge claims to the prize, which would be paid by the British Parliament. A solution to the problem required positions of the stars to 1' or better and, for the method of lunar distances, positions of the Moon to 1'. Also, one required accurate clocks. Later (1755) the German astronomer Tobias Mayer produced tables of the Moon's position good enough for the geographical longitude to be

Fig. 4.11. James Bradley (1693–1762), third Astronomer Royal. Photo courtesy of The Royal Society, London.

determined to 1°. (His solar and lunar tables were used as the basis of the *Nautical Almanac*.) In 1765 Mayer's widow was awarded £3000 by the Board of Longitude for her husband's efforts, but the lion's share of prize money went to the chronometer maker John Harrison, who built the first useable marine chronometer. Harrison received £10 000 by 1765, primarily for his chronometer Number 4 (known as H4), and an additional £10 000 in 1772.[34] The requirement for the prize-winning timekeeper was that it had to be accurate to 2 minutes of time over the course of a trans-Atlantic voyage, and all details of its construction had to be turned over to the Board of Longitude. Due to the accuracy of the star catalogues produced at Greenwich, the tables of positions of the Moon, Sun, and planets (finding their place in the *Nautical Almanac*), and the improvements in chronometers, the problem of longitude at sea was essentially solved by the end of the eighteenth century. The Board of Longitude, deemed no longer necessary, was dissolved in 1828.

The third Astronomer Royal was James Bradley (Fig. 4.11), easily regarded

Fig. 4.12. Bradley's $12\frac{1}{2}$ foot zenith sector, with which he discovered the aberration of light. Photograph courtesy of the National Maritime Museum, London.

as one of the greatest observational astronomers of all time.[35] Working with Samuel Molyneux in 1725 Bradley was able to show that the declination of the star γ Draconis was not constant, confirming a previous (1669) result of Robert Hooke. Using a $12\frac{1}{2}$ foot zenith sector (Fig. 4.12) set up at his aunt's house in Wanstead (Essex), which could be pointed up to $6\frac{1}{4}°$ from the zenith and which could be used to observe a greater number of stars than Molyneux's instrument, Bradley found that the variation of declination was proportional to the trigonometric sine of the celestial latitude of the stars and that the period of variation was 1 year. (The celestial latitude is the number of degrees an object is north or south of the ecliptic.) Actually, since γ Draconis is at a right ascension of about 18 hours, the direction of increasing declination is the same as the direction of increasing celestial latitude. The shift in the celestial latitude is found to be

$$\Delta\beta = -k \sin \beta \sin (\odot - \lambda_*),$$

where k is the aberrational constant (given below), β is the celestial latitude, and $\odot - \lambda_*$ is the difference in the celestial longitude of the Sun and the star. I give here only one component of the apparent motion of the star about its unaberrated position; the path on the sky around the mean position is an ellipse with the true position at the center; the major axis is parallel to the ecliptic and the semi-major axis is equal in size to the aberrational constant. In the case of γ Draconis (and other stars as well) Bradley found that a star was at its most northerly position (maximum $+ \Delta\beta$) when it culminated at 18.00 h local time ($\odot - \lambda_* = -90°$), and it was at its most southerly position (maximum $- \Delta\beta$) when it culminated at 06.00 h local time ($\odot - \lambda_* = +90°$).

The variation about the mean position could not be due to annual parallax for two reasons. First, the variations were the same for all stars at the same celestial latitude, and it was unlikely that all stars were at the same distance. The apparent path on the sky due to parallax is another ellipse (superimposed upon the variations due to aberration) which has the same shape as the aberrational ellipse for the same celestial latitude, but the size of the semi-major axis of the parallactic ellipse is equal to the trigonometric parallax. The shift in celestial latitude due to annual parallax is given by

$$\Delta\beta = -P \sin \beta \cos (\odot - \lambda_*),$$

where P is the trigonometric parallax in arc seconds, and β and $\odot - \lambda_*$ are the same as given above. Thus, the maximum northerly position of a star due to parallax would occur when the Sun's longitude was opposite that of the star, whereas the most northerly shift due to the aberration effect would occur when the Sun's longitude was 90° less than the longitude of the star. My second point is that the shifts due to parallax and aberration are rotated by 3 months of time with respect to each other.[36]

Bradley correctly concluded that the observed variations of stellar declination were due to the Earth's motion and the finite speed of light. He presented his results to the Royal Society in January of 1729. Whereas Roemer obtained a value of 11 minutes for the light-travel time from the Sun to the Earth, Bradley's aberration constant of $20''.2$ leads to a value of $8^m 12^s$, very close to the accepted modern value. (Today we can differentiate between annual aberration, due to the Earth's orbit around the Sun, and diurnal, or daily, aberration, due to the Earth's rotation on its axis, with a maximum value of about $0''.3$.)

Bradley's other great discovery was the nutation (or wobbling) of the Earth's axis. As we saw in Chapter 1, the Greeks already knew about the precession of the equinoxes. This is a uniform motion of the vernal equinox westward at a rate of $50''.3$ per year. Newton had predicted that the Moon's motion would give rise to a variable perturbation in the celestial longitudes of the stars and also in the obliquity of the ecliptic. This is due to the Moon's orbital plane being inclined to the Earth's orbital plane (the ecliptic) by an angle of $5° 09'$. It turns out that the obliquity of the ecliptic varies with a period of 18.6 years. This causes the celestial latitudes of the stars to have a periodic component on that time scale. From Bradley's observations at Wanstead from 1727 to 1747 with the $12\frac{1}{2}$ foot zenith sector (covering more than a nutational period) he was able to demonstrate the principal nutational term, with a magnitude of about $9''$. (The nutation is really the sum of many perturbations superimposed on the effect of precession, the largest term of which was discovered by Bradley.)[37]

Bradley was the third, but not the last, of the Astronomers Royal who, upon assuming that office, faced a situation in which the instrumentation at Greenwich was in dire need of improvement. He was granted £1000 for the purchase of new equipment. He ordered an 8 foot brass mural quadrant and an 8 foot transit instrument, both by John Bird. This transit instrument defined the Greenwich meridian from 1750 to 1816. A new observatory building was completed in 1750 to the south of the old observatory, where these instruments and Bradley's own $12\frac{1}{2}$ foot zenith sector were installed. Greenwich Observatory, after 75 years, had really come of age.

> From the year 1750 may be dated the commencement of a series of observations which in point of accuracy may bear a comparison with those of modern times. Henceforward the records of Greenwich Observatory embody a collection of materials, which have almost

exclusively formed the groundwork of every investigation undertaken in
modern times, for the purpose of improving the solar, lunar, or planetary
tables.[38]

This was not due just to having better-constructed instruments of larger
size (able to measure smaller angles). The raw data were only judged
complete if accompanied by measurable alignment and collimation errors of
the telescope, errors in the clock used, and measurements of the temperature
and barometric pressure (for the elimination of the effect of refraction). Due
to the discovery of aberration and nutation, these effects could be tabulated
and subtracted from the observations. Thus it was that the accuracy of
declination measurements, for example, had increased 10-fold since 1690,
and 60-fold since the time of Tycho.[39]

Though Bradley's observations of the Sun, Moon, planets and stars were
not published until long after his death (one volume in 1798, a second in
1805), they provided the raw material for the development of the study of
astronomical errors, as derived by such greats as Bessel, who reduced
Bradley's stellar observations made at Greenwich from 1750 to 1762 and
published a catalogue of 3222 stars for the epoch 1755 in his classic
Fundamenta Astronomiae (1818). The uncertainty of the derived declinations
was 4″, while for right ascensions the uncertainty was 1 second of time. Such
was the interest in positional astronomy, and in Bradley's observations in
particular, that Arthur Auwers undertook a complete rereduction of
Bradley's data from 1865 to 1883. Bradley's positional data gave good 'first
epoch' positions for the subsequent determination of proper motions of stars
(their angular motions across the line of sight with respect to a distant frame
of reference) which were instrumental in Kapteyn's 1904 discovery of star
streams in the galaxy. Given the number of astronomers who had worked in
that intervening century and a half and the improvement in observational
methods, it is incredible that Bradley's observations were still regarded as a
fundamental source of data.[40]

The activity at Greenwich from 1750 until 1835 under Bradley, Bliss,
Maskelyne and Pond remained pretty much along the lines already drawn.
The astronomers were involved in whatever could be done to improve
navigation at sea, from the compilation of still more accurate star catalogues
and ephemerides, to the calibration of chronometers for the Royal Navy. We
have already mentioned the publication of the *Nautical Almanac*, which was
primarily the result of the efforts of Maskelyne. British astronomers,
including those at Greenwich, also became involved in the expeditions of
1761 and 1769 to observe the transits of Venus (to help determine the scale
of the solar system). Thus, more research-oriented areas of positional

Fig. 4.13. William Herschel (1738–1822). From Herschel, Mrs John, *Memoir and Correspondence of Caroline Herschel*, New York: D. Appleton, 1876, facing p. 118.

astronomy were being pursued. But, as described by John Herschel, concerning the astronomy carried out in Britain at the end of the eighteenth century and the beginning of the nineteenth century, 'The chilling torpor of routine had begun to spread itself over all the branches of science which wanted the excitement of experimental research.'[41] To measure the positions of stars restricted one to a very small slice of the heavens at any given time. There were others, like William Herschel and his son John, who had wanted to investigate the whole pie, not just the individual crumbs.

The story of William Herschel (Fig. 4.13) has been written many times, and it was stated in the Preface that concern is focused on institutions of research, not the whereabouts of a given individual, unless there were no institutions at all of note. Yet, 'the rise of Herschel was the one conspicuous anomaly in the astronomical history of the eighteenth century. It proved decisive of the course of events of the nineteenth'.[42]

At first a musician, Herschel became passionately addicted to scanning the heavens and became famous as a result of his serendipitous discovery of the

Fig. 4.14. William Herschel's 40 foot reflector, which contained a 48 inch diameter mirror made of speculum metal. From Caroline Herschel's memoirs, facing p. 29.

planet Uranus on 13 March 1781. As Herschel himself says:

> It has generally been supposed that it was a lucky accident that brought this star to my view. This is an evident mistake. In the regular manner [that] I examined every star of the heavens, not only of that magnitude but many far inferior, it was that night *its turn* to be discovered.[43]

The object was first thought to be a comet, but when its orbit was calculated, it proved to be the first new planet known since ancient times. Herschel was granted a Royal pension of £200 a year and eventually settled at Slough, near Windsor Castle, where his only official task was to show the Royal family the celestial wonders from time to time. This salary allowed Herschel to devote himself full time to astronomy, but it would have been insufficient to live on without his other income derived from the sale of telescopes, amounting over the course of time to a total of £16 000 (gross, not net). The construction of telescopes for sale and personal use coupled with the discoveries he made has established Herschel's reputation as one of the greatest astronomers of all time. He clearly recognized that the faintest objects could only be observed with a telescope of maximum light-gathering power; hence, the area of the objective had to be made as large as possible. The limit to the size of refractors at that time was about 6 inches. Herschel constructed reflecting telescopes with objectives of speculum metal. His largest (Fig. 4.14), with a diameter of 48 inches, was the largest telescope ever until a 72 inch was constructed by William Parsons, Third Earl of Rosse, in 1845. Though the speculum mirrors tarnished rapidly and needed frequent refiguring, their 100-fold increase of light-gathering power rendered visible objects 5 magnitudes fainter. One could penetrate 10 times further into space.

Herschel is well known for his surveys of the heavens. They were just that – an answer to the questions: 'What types of objects are there in the universe? What can be said about their arrangement, even their evolution?'. Herschel's first survey of the sky (1779) included all objects to fourth magnitude visible at his latitude in England. His second survey, carried out with a reflector of 7 foot focal length and completed in 1782 (during which he discovered Uranus), penetrated to 8th magnitude. His third, and most complete, survey (1783–1802) brought the number of known nebulae to 2500. His study of nebulae led Herschel to speculate on the evolution of stellar systems under the principle that, by having a snapshot of many nebulae of a certain kind at different stages of evolution, it was possible to make statements concerning how an individual nebula evolves. Herschel's observations of planets and their satellites along with his determinations of planetary rotational periods remained the definitive studies for many years. He founded the study of double stars, having discovered nearly a thousand examples. (At first he had hoped to use the chance pairing of stars to determine the trigonometric parallax of the brighter, presumably nearer, stars, but he found that the double stars were not chance pairings at all; the components of a double star pair orbited each other under the mutual attraction of gravity.)

Herschel's 'star gauges' allowed him to construct the first model of the Milky Way based on observational evidence. However, he had few immediate successors in the field of stellar statistics; the most important were his son John, who executed similar projects in the southern hemisphere (Chapter 8) and Wilhelm Struve, whose 1847 book on stellar astronomy (Fig. 4.15) provided the first observational evidence of an interstellar absorbing medium. By the beginning of the twentieth century, stellar statistics was a primary field of research. One might say that the work of William Herschel

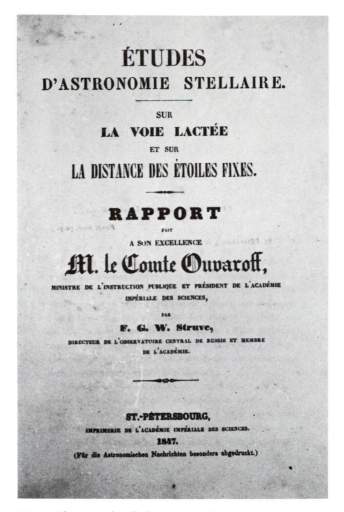

Fig. 4.15. Title page of Wilhelm Struve's book on stellar astronomy.

touched on and improved every aspect of astronomy except the type of positional work carried on at Greenwich.[44,45]

The story of William Herschel is one of many, in the nineteenth century in particular, of an advanced amateur under no pressure by committees or funding agencies to pursue a particular line of research. Such an individual could do first-class work because he could still afford to buy his own first-class equipment. Today a 'state-of-the-art' telescope typically costs millions of dollars and has a staff of perhaps 50–100 to keep it running and use it. Yet there will always be room for trained scientists at the 'fringe' of a field who can provide insight into new lines of research, precisely because they are not

Fig. 4.16. George B. Airy (1801–1892), seventh Astronomer Royal. Photo courtesy of the National Maritime Museum, London.

saturated with the 'established' notions of what can and can not be done, and what is assumed to have been solved with sufficient clarity.

Returning to the Royal Observatory, we have already made reference to the methodical reign of George Biddell Airy (Fig. 4.16), who incidentally was the last Astronomer Royal to receive a Royal Warrant.[46] Airy is to be considered one of the most important British astronomers of the nineteenth century, serving as Astronomer Royal from 1835 to 1881. The prime accomplishment of his reign was making the Greenwich observations readily available soon after they were made. Whereas the first three Astronomers Royal regarded the observations made at Greenwich to be their personal property,[47] and the next three still did not press for prompt publication of Greenwich data, it was Airy who saw to it that the backlog of observations was cleared up and that the future observations would be more quickly reduced and disseminated. Airy, being very much a Victorian moralist, related to the concept of duty, and he regarded it to be the duty of the Royal Observatory to carry on its tradition of positional astronomy, concentrating on observations of the Moon, Sun, planets and catalogues of stars (in that order), as befitting the original purpose of the observatory, and only secondarily pursuing projects of stellar astronomy (in the style of William Herschel) or, later, astrophysics (spectroscopy, photography, solar astronomy).

For observations of the Moon (the highest priority object at Greenwich) Airy designed and procured an altazimuth. The engineering was carried out by Ransome and May of Ipswich, with the optics and instrumentation by Troughton and Simms of London. Used from 1847 to 1897, Airy's altazimuth allowed observations of the Moon to be made other than on the meridian so that the Moon's position could be measured at night for any phase, thus allowing all parts of its orbit to be covered. Tables of the Moon then deviated up to 40″ from the observed positions. Greenwich observations on this altazimuth led to a refinement of lunar theory and observation.

Perhaps the most famous instrument from Airy's day was the new transit circle (Fig. 4.17), constructed by the same firms which built the altazimuth. The Greenwich meridian had been defined by Bradley's 8 foot transit instrument until 1816, when it was replaced by a 10 foot transit instrument built by Troughton. Airy's transit circle was installed in 1850 and was located 19 feet to the east of the pier of the previous two transit instruments. This makes a difference of $\frac{1}{50}$ second in the local apparent time, insignificant then, but not so today. The Airy transit circle has defined the Greenwich meridian since its installation; its position has defined the world's Prime Meridian since 1884. More than three quarters of a million observations were made on it in more than a century of use.[48]

The third principal instrument installed during Airy's regime was a 12.8 inch visual refractor by Merz of Munich (1859). Though Airy had a preference for meridional astronomy, there was an increased interest in extrameridional observations of comets, double stars and the development of astrophysical methods. The 'Great Equatorial' paved the way for the 28 inch visual refractor (1893), the 26 inch photographic refractor and the 30 inch photographic reflector (1896; which were mounted on the same pier),

Fig. 4.17. The Airy transit circle, defining the world's prime meridian since 1884. Photograph courtesy of the National Maritime Museum, London.

and the 36 inch Yapp reflector (1932). Since the Second World War we have seen the fulfilment of William Herschel's call for more light-gathering power in the construction of the 100 inch Isaac Newton Telescope (originally a 98 inch) and the 4.2 meter William Herschel Telescope (Chapter 8), located in the Canary Islands but managed by RGO.

It is not possible to describe here all the projects carried out at Greenwich

Fig. 4.18. Domes of the equatorial group at Royal Greenwich Observatory, Herstmonceux. Photograph courtesy of the Royal Greenwich Observatory, Herstmonceux.

in the last 150 years. Of the most significant are the collaboration in observations of the Venus transits of 1874 and 1882 and observations of the minor planet Eros in 1931, which led to the modern value of the distance to the Sun.[49] Greenwich Observatory could be proud to be the first observatory to complete its assigned portion of the *Carte du Ciel* project, using the adopted standard photographic refractor (33 cm, f/10.4) like that of the Henrys at Paris Observatory. They finished the mapping and catalogue for their assigned part of the sky (declination $+90°$ to $+65°$) in 1909.[50] The collaboration of Greenwich Observatory in the 1919 solar eclipse expedition demonstrated the gravitational bending of light predicted by Einstein's General Theory of Relativity. The object glass of the *Carte du Ciel* refractor was used in Brazil for photographs of this famous eclipse.

The activity at Greenwich Observatory in the past 50 years has undergone some major changes, the most significant being the location where the observations were carried out. Due to the urbanization near the observatory (even in Airy's time) the seeing conditions at Greenwich had begun to deteriorate. After the Second World War the Admiralty purchased Herstmonceux Castle in rural Sussex, and the move to the castle was carried out under Spencer Jones and Woolley, being completed in 1958. This entailed a change of name as well. What had previously been the Royal Observatory, Greenwich, became the Royal Greenwich Observatory at Herstmonceux. This move to the Sussex countryside certainly improved the possibilities of carrying out observations under better skies, but in retrospect, once the 98-

inch Isaac Newton Telescope (INT) was finished in 1967 after 20 years, it was unfortunate to have been situated in Britain at all, for even at Herstmonceux the seeing was not good by world standards. By 1984 the INT was operating at La Palma in the Canary Islands. There it is part of the Roque de los Muchachos Observatory, a joint venture of Spain, the United Kingdom, The Netherlands, Denmark and Sweden.

The story of Greenwich Observatory in the second half of its existence manifests at the start the individual style of the Astronomer Royal, but as time goes on it becomes the story of teamwork. Today the Astronomer Royal is no longer required to reside at RGO. The observatory is guided by a director. It is no longer administered by the Admiralty or Ministry of Defence, since 1965 being subordinated to the Science and Engineering Research Council (as it is now called). Since 1675 Greenwich Observatory has evolved from a one-man show dedicated to finding out 'the much-desired longitude of places for the perfecting the art of navigation' to a large research organization still involved with positional astronomy but also dedicated to astrophysics. Many of the old telescopes are preserved at the Old Royal Observatory. The large number of medium-sized instruments at Herstmonceux are gradually being phased out, but the operation of the large reflectors in the Canary Islands ensures that the RGO is still involved with state-of-the-art astronomical research.*

* In 1986 the Science and Engineering Research Council decided that RGO would move again. Herstmonceux Castle would be sold and the observatory would be based in Cambridge.

5

*Pulkovo Observatory**

L'Astronomie, par la sublimité de son sujet, occupe une place
éminente parmi les sciences naturelles. Elle est, par excellence, la
science naturelle exacte.
Wilhelm Struve (1845)

During the seventeenth century Russia's only seaport was Arkhangelsk,
near the Arctic Circle on the shores of the White Sea. Then, in 1703, the
Russian Tsar Peter the Great commanded that a fortress be built at the
mouth of the Neva River, which empties into the Baltic, to defend the land he
had won in battle from the Swedes. Tens of thousands of men died
transforming the marshy islands of the Neva into one of the most elegant
cities on earth, St Petersburg (now Leningrad). By the time Peter the Great
died, the city of Petersburg, crossed as it was by canals and filled with palatial
architecture, reminded one of Amsterdam or an immense version of
Venice.[1,2]

About the time of Peter's death in 1725 the Russian Imperial Academy of
Sciences was founded. Near the fortress which first marked the city the
Academy built the *Kunstkammer*, or 'Cabinet of Curios' (a natural history
museum), which contained scientific instruments, fossils, minerals,
examples of taxidermy from the not so unusual (a zebra) to the bizarre ('the
skin of a Frenchman tanned and stuffed'),[3] and in its tower also sported
an observatory. Delisle, a well-qualified Frenchman, was invited to St
Petersburg as the first director. He occupied that post until 1747 when the
building housing the observatory burned down. It was rebuilt and re-
equipped with the same type of simple instruments, but in 1761 the Russian
government decided to promote more accurate observations and purchased
a mural quadrant of 8 foot radius and a transit instrument of 5 foot focal
length, both by Bird of England.[4] However, like the Paris Observatory, the
observatory at St Petersburg was designed more for show than for
astronomical observing, and its placement in the heart of the city was no

---

* The Russian style spelling is "Pulkovo"; the French style is "Poulkova"; the German
  style is "Pulkowa".

Fig. 5.1. Dorpat Observatory, at which Wilhelm Struve carried out double star astronomy and measured the parallax of Vega. From Herrmann, Dieter B., *The history of astronomy from Herschel to Hertzsprung*, Cambridge University Press, 1984, Fig. 70, p. 181.

advantage either. There were regular discussions concerning the establishment of an observatory which would be well equipped and well situated, but nothing materialized until the late 1820s.

The precursor to Pulkovo Observatory was not the Academy's observatory but rather the observatory at Dorpat (now Tartu in Estonia; Fig. 5.1). From 1817 its director had been Friedrich Georg Wilhelm Struve (Fig. 5.2), who had obtained his doctorate at Dorpat University in 1813 at the age of 20. There he taught courses in astronomy and geodesy, upgraded the astronomical instrumentation (including ordering a 9.6 inch refractor from Fraunhofer – then the largest refracting telescope in the world), measured double stars, and made accurate micrometric measurements of stellar positions. The last-mentioned led to his greatest discovery – the determination of annual parallax of the star Vega. Along with Friedrich Wilhelm Bessel and Thomas Henderson he became one of the first three people to produce this final proof of Copernicanism. Throughout his life W. Struve was also involved in geodesy. His greatest project in this regard was the Russian–Scandinavian measurement of $25° \, 20'$ in latitude, finished in 1855.[5]

In 1827 his *Catalogus novus generalis stellarum duplicium et multiplicium* was published. It was a catalogue of measurements of double and multiple

Fig. 5.2. Wilhelm Struve (1793–1864), first director of Pulkovo Observatory. From A.N. Dadaev, *Pulkovskaia Observatoriĩa*, Leningrad, 1972, p. 11.

stars made with the Dorpat 9.6 inch refractor, most of which he had discovered. For the work that went into this catalogue Struve was awarded the Gold Medal of the Royal Astronomical Society.

The same year as the publication of Struve's double star catalogue, his colleague and former professor at Dorpat, the physicist Georg Friedrich Parrot, was called to St Petersburg to serve as a member of the Academy of Sciences. Parrot was called upon to put together a plan for a new major observatory. Because of Struve's astronomical and geodetical work and his friendship with Parrot, Struve could not fail to win the recognition of the Academy and even the Tsar himself. In January of 1831,[6] having just returned from abroad, he was honored with a personal interview with Tsar Nicholas I. Let us give Struve's own account of the interview, at which the Minister of Public Instruction, Prince von Lieven, was also present:

> Having listened to my report on the late scientific journey, and after having graciously granted an increased sum to the observatory at Dorpat, the Emperor condescended to put to me the following question:

> 'What is your opinion of the observatory of St. Petersburg?'

> I did not hesitate to respond, in all frankness and in accordance with the exact truth, that the observatory of the Academy did not at all correspond to the present demands of science, and that it partook of the nature of all the establishments of its kind placed in the midst of large cities, as those of Vienna, of Berlin, etc., and even of Paris, where the meridian instruments ought to be removed from the colossal edifice constructed under the reign of Louis XIV, and be placed in modest apartments adjacent to the principal structure.

> Having listened to this reply, his Majesty addressed the minister of public instruction, saying that he regarded the establishment of an observatory of the first rank near to the capital as an object of high utility and importance to the scientific honor of Russia.

> The minister did not fail to inform the Emperor that the Academy had for some years occupied itself with the project of a new observatory, and that he had only awaited the completion of the plans and drawings in order to lay them before his Majesty. Then the Emperor ordered that the project should be presented to him as soon as it should have been matured. Finally, his Majesty condescended to direct his attention to the choice of the location for the institution to be erected. The minister having mentioned the site to the north of the city and offered as a gift to the Academy, the Emperor condescended to express himself in the following terms:

> 'How? The Academy thinks to place the new observatory quite near

the city on the north side, and upon a sandy and marshy soil? That is hardly advisable. I would suggest another position. It is upon the heights of Poulkova that the observatory should be placed.'

Then his Majesty condescended to address to me the following words:

'Sir Astronomer, you perhaps think it strange that the Emperor should wish to correct the Academy in a scientific matter. But do you know Poulkova, and what do you think of the site?'

My reply was that in 1828, passing for the first time by Poulkova in the company of the Baron von Wrangell, I had been so struck with its position that I had, as if prophetically, exclaimed: 'There upon the hill of Poulkova it is that we shall one day behold the observatory of St Petersburg'.[7]

Matters progressed all too slowly for Parrot, but on 28 October 1833 (O.S.) the Minister of Public Instruction (now S. S. Uvarov) was authorized by the Tsar to begin building the observatory. On the 31st of October Uvarov created a new commission under the chairmanship of Admiral A. S. Greig and consisting of the academicians V. K. Wisniewski, G. F. Parrot, F. G. W. Struve, and P. N. Fuss. (After continuing to express impatience concerning the speed at which things were progressing, within a year Parrot was replaced by E. KH. Lenz.) In November of 1833 the commission presented Uvarov with a report of their site survey within the vicinity of St Petersburg and recommended that the new observatory be located at Pulkova (19 km south of the city on a hill 75 m above sea level; Fig. 5.3).

Money was allocated for the purchase of instruments. Having been confirmed in April of 1834 as the future director of the observatory, W. Struve went abroad from June through October of 1834 to visit other European observatories, seek the advice of other astronomers, and order instruments. To Fuss he wrote:

> ... the founding of Pulkovo Observatory is regarded abroad as an event unparalleled in the history of science, and therefore it excites interest far greater than everything I myself might be able to muster. In particular, this interest was especially keen in Berlin, since there they are occupied with the construction of a new observatory ... my visit to the Berlin Observatory, close to being finished, did not make me want to change the least little thing in our plans ...[8]

The primary instruments (Table 5.1) were: a transit instrument (for the determination of absolute right ascensions) and vertical circle (for the determination of absolute declinations) by Ertel; a meridian circle (for determining differential stellar coordinates) and transit instrument (for

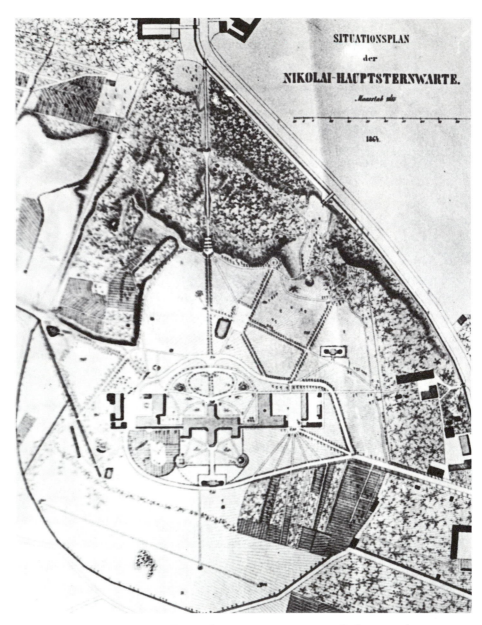

Fig. 5.3. Site plan for Pulkovo Observatory. From Otto Wilhelm Struve's 1865 history of the observatory.

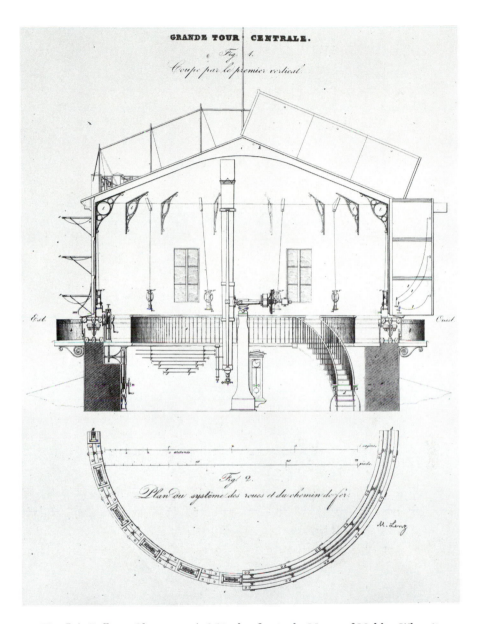

Fig. 5.4. Pulkovo Observatory's 15 inch refractor by Merz and Mahler. When it went into operation in 1839, it was the largest telescope in the world. A twin of this telescope was made for Harvard College Observatory. From W. Struve's 1845 *Description* of the observatory. Photograph courtesy of Alan H. Batten.

Table 5.1. *The primary instruments at Pulkovo Observatory at the time of its founding* (1839)[9,10,11]

Instrument	Manufacturer	Place manufactured	Objective diameter	Focal length	Magnification		Field size		Place situated
					Min	Max	Max	Min	
Transit instrument	Ertel	Munich	5".85	102"	95	292			E end of W room
Vertical circle	Ertel	Munich	5".90	77"	77	215			W end of W room
Meridian circle	Repsold	Hamburg	5".81	83".2	170	245			E room
Transit instrument on prime vertical	Repsold	Hamburg	6".25	91"		270			S wing
Large refractor	Merz & Mahler	Munich	14".93	270".6	152	1822	13'.2	1'.1	Central dome
Heliometer	Merz & Mahler	Munich	7".4	123"	79	316	27.3	7.1	E dome
Small refractor	Plössl	Vienna	6".8						W dome
Comet seeker	Merz & Mahler	Munich	3".8	[30"]	18	86	3°35'	18'	portable
Comet seeker	Plössl	Vienna	3".7		17	25	4°26'	2°9'	portable

DESCRIPTION

DE

L'OBSERVATOIRE ASTRONOMIQUE

CENTRAL

DE

POULKOVA.

PAR

F. G. W. STRUVE,

MEMBRE DE L'ACADÉMIE IMPÉRIALE DES SCIENCES DE St-PÉTERSBOURG, PREMIER ASTRONOME ET DIRECTEUR
DE L'OBSERVATOIRE CENTRAL.

St.-PÉTERSBOURG, 1845.

=

IMPRIMERIE DE L'ACADÉMIE IMPÉRIALE DES SCIENCES.

Fig. 5.5. Title page of W. Struve's two-volume work on the instruments and plans of research of Pulkovo Observatory. Concerning this work, David Gill, fifth Astronomer Royal at the Cape of Good Hope, said, 'There is inspiration to be found in nearly every page of it, for its author had the true genius and spirit of the practical astronomer – the love and precise methods of observation and the inventive mechanical and engineering capacity'. Photograph courtesy of Alan H. Batten.

determining the astronomical constants of aberration and nutation; to be fixed on the prime vertical, the plane perpendicular to the meridian) by Repsold; a 15 inch refractor by Merz and Mahler, which would again give Struve access to the world's largest refractor; and a heliometer, also by Merz and Mahler, for the determination of stellar parallaxes. He also ordered a smaller (7 inch) refractor from Plössl and two comet seekers. For keeping time he ordered pendulum clocks by Hauth in Petersburg, Kessels in Altona, Muston in London, Arnold & Dent in London, and Tiede in Berlin. Three chronometers were to be built by Kessels, three more by Hauth, and one by Arnold & Dent, along with five alarm clocks by Hauth. They would need portable instruments for geodesy, meteorological equipment, and equipment for the Pulkovo mechanical shop, which Struve ordered as well.

The construction of the observatory proceeded under the direction of its architect Aleksandr Briullov. The cornerstone was laid on 21 June (3 July) 1835 and the building was completed in 1839. The main building (as it appears today after its postwar reconstruction) consisted of two stone buildings, a west one and an east one, and a central observational part united by corridors to the east and west buildings. At the top of the Pulkovo skyline one saw three octahedral towers with conical domes. In the middle dome was the 15 inch refractor (Fig. 5.4). In the east dome was the heliometer, and in the west dome the 7 inch refractor. The Ertel transit instrument and vertical circle were to be found in the west observing room (below and west of the large refractor), while the Repsold meridian circle was in the east room and the Repsold prime vertical transit was in the south wing.

The observatory was inaugurated on 7(19) August 1839. Its statutes defined its purpose to be:

1) to provide uninterrupted observations as accurate as possible which serve the progress of astronomy;
2) to carry out corresponding observations for geographical undertakings within the empire and for carrying out scientific excursions; and
3) to provide all measurements for the improvement of practical astronomy in its application to geography and navigation, and provide the opportunity for practical exercises in geographical determinations of position.[12]

When all was accounted for by the time the observatory opened, some 2 100 000 rubles in currency (600 000 silver rubles) was spent on the observatory (1 700 000 was allocated for the buildings, while 300 000 rubles in currency went for the construction and transport of the instru-

Fig. 5.6. Otto Wilhelm Struve (1819–1905), second director of Pulkovo Observatory. The most respected Russian astronomer of the second half of the nineteenth century. From Dadaev, 1972, p. 19.

ments).[13] It was thus as lavishly designed and executed as Tycho's observatory.

The early history of the observatory and an extremely detailed account of the instrumentation is to be found in W. Struve's classic two-volume *Description de l'Observatoire astronomique central de Poulkova*, published in 1845 (Fig. 5.5).[14] All subsequent accounts of the history of Pulkovo derive from this work.[15–20] Because of its wealth of detail, Struve's book served as a model for the construction and outfittings of many other observatories of the nineteenth century. For example, the Harvard 15 inch refractor (p. 125), also by Merz and Mahler, was known as the Pulkovo twin.

Pulkovo (like Greenwich was founded for practical astronomical reasons. Its reputation was to be based on the quality of astronomical measurements of the positions of celestial bodies. To produce such data, it was not enough just to have the largest and finest instruments made by German manufacturers. Beginning with the work of Gauss, Bessel, and W. Struve himself in the early nineteenth century, one also had to apply the 'science of observation', whereby the systematic errors of the instruments and observers are taken into account after the observations are made, while sets of data are used under the guise of a mathematical theory of errors and a thorough understanding of the engineering of the equipment to produce the 'most likely' results. One contemporary of Struve wrote:

> Instruments no smaller than ours are situated at the English observatories, but their shifts present such difficulties that the local astronomers can hardly decide what to do; indeed, the calibration of an instrument with the help of a level is carried out for the most part once a week, and a test reading on the tube by means of a distant reference is not always possible. At Pulkovo such calibrations are repeated every day up to several times, and result in observations of such accuracy as to be unattainable with other instruments.[21]

Though the Astronomer Royal at the time, George Airy, might beg to differ with the above-quoted author, no less an authority than the American astronomer Simon Newcomb considered the Pulkovo observations of such high quality that 'one observation on the Pulkovo vertical circle was equivalent to twenty, thirty, or even forty observations made by average observers on a meridian circle'.[22]

As explained in Chapter 4, in order to produce accurate star positions one must reduce raw observations to absolute positions by correcting for systematic errors of the instrument and observer, and then eliminate the effect on the observations of their being made from a spinning, wobbling platform (the Earth) revolving around the Sun. One must account for aberration, precession, nutation and also the effect of atmospheric refraction. From the observations at Dorpat, W. Struve felt he had determined the refraction corrections, though the determination of refraction is still a problem. (The fourth edition of the Pulkovo refraction tables was published in 1956). The customary value for nutation ($9''.210$), which we use by international agreements of 1896 and 1964 is very close to the value published by Pulkovo astronomer C. A. F. Peters in 1842—$9''.214$. For aberration W. Struve used his own observations at the Pulkovo prime vertical transit made during the years 1840–2. He obtained a value of $20''.471 \pm 0''.0135$.[23] (This was one of his last large observational programs, and after 1845 he more and more left the administration of the observatory to his son, Otto Wilhelm Struve.) Until the work at Pulkovo the

Fig. 5.7. Simon Newcomb (1835–1909), of the United States Naval Observatory and the Nautical Almanac Office. He was the most widely respected American astronomer of the second half of the nineteenth century and also a good friend of his counterpart at Pulkovo, O. W. Struve. Photograph courtesy of the US Naval Observatory.

best determination of aberration was that of Bessel (20″.475). In 1896 Newcomb suggested that the value of 20″.470 be adopted. However, the average of its values for all determinations at Pulkovo from 1840 to 1880 (20″.493), also known to Newcomb, was almost in agreement with the value adopted in 1964 at the Hamburg meeting of the International Astronomical Union – 20″.4958. That value obtained at Pulkovo over the years 1953 to 1959 (20″.4965) must have figured prominently in the decision of the IAU.[24] The differences we are discussing here amount to less than a 100-millionth of a degree!

The determination of precession was left to Otto Wilhelm Struve (Fig. 5.6). It earned him a master's degree from the University of St Petersburg in 1841 and the Gold Medal of the Royal Astronomical Society (RAS) in 1850. His value (50″.235/year) was used throughout the world for 50 years.[25]

Thus it was that the fundamental constants of positional astronomy, required for the production of star catalogues of the highest precision, were

Fig. 5.8. The US Naval Observatory's 26 inch refractor, shortly after its completion in 1873. Newcomb is at the eyepiece. Admiral Benjamin Sands is the man with white hair. Photograph courtesy of the US Naval Observatory.

all determined very accurately at Pulkovo. Consequently, their star catalogues are also of high quality. They have produced such catalogues of right ascensions and declinations of stars for the epochs 1845, 1865, 1885, 1905, 1930 and 1955. The first catalogue contained 374 stars down to the fourth magnitude, while in the last there were 1046 bright and faint stars.

Fig. 5.9. The US Naval Observatory in the 1880s, before the 1893 move to its present location. The dome of the 26 inch refractor is behind the tree, to the right of center. Photograph courtesy of the US Naval Observatory.

It was the fulfilment of the goal of producing positional measurements of the greatest possible accuracy which led to the assessment of Pulkovo Observatory as the 'astronomical capital of the world'.[26] Other positional work carried out elsewhere was equally important, however. While fundamental catalogues such as Pulkovo's contained data as accurate as possible on a relatively small number of usually bright stars, catalogues of positions of as many stars as possible are also important. The most important project in this regard was the *Bonner Durchmusterung* (BD). It contained the approximate positions and brightnesses for 324 198 stars brighter than tenth magnitude and north of declination $-2°$. This was carried out by F. Argelander and E. Schönfeld at the Bonn Observatory during the years 1852–9. Schönfeld later completed a southern BD, containing 133 659 stars in the declination range $-23°$ to $-2°$. Finally, observers in Argentina produced the *Cordoba Durchmusterung* (CD), containing 613 953 stars in the rest of the southern sky. Thus over a million stars were catalogued, and with instruments as modest as a 3 inch refractor. These catalogues prove invaluable for studies of stellar positions and luminosities (i.e. studies of variable stars), and many stars are still referred to by their BD (or CD) numbers.[27]

Positional astronomy, of an observational, theoretical, and computational nature, was being carried on in America, most notably at the United States

Naval Observatory (USNO). The most eminent American astronomer in the second half of the nineteenth century was Simon Newcomb (Fig. 5.7), who had become a staff member of the Naval Observatory in 1861. He directed the construction of the Naval Observatory's 26 inch refractor (Fig. 5.8), completed in 1873. Four years later he became superintendent of the Nautical Almanac Office, based at the USNO, which is responsible, along with HM Nautical Almanac Office, for producing the annual volume of ephemerides for the Sun, Moon, and planets used by all astronomers. Until recently the definitive models of the movements of the solar system bodies were primarily due to the efforts of Newcomb. And it would not surprise the reader to learn that Newcomb and his counterpart in Russia, Otto Wilhelm Struve, were good friends. The Newcomb archives at the Library of Congress contains many letters of O. W. Struve, extending over a period of 37 years. Many astronomical topics are discussed, among them the transits of Venus of 1874 and 1882, the reality of the companion of Procyon,[28] measurements of double stars, the construction of telescopes, possible changes in the brightness and form of the Orion Nebula, and the politics of astronomy.

How long Pulkovo would remain the 'astronomical capital of the world' depended essentially on the state of their equipment in comparison to other observatories and the quality and progressiveness of their work, taking into account developments in other areas of science.

In 1862 O. W. Struve succeeded his father as Pulkovo Observatory director (see Table 5.2 for a list of the Pulkovo directors). By this time Pulkovo no longer had the largest telescope in the world. The firm of Alvan Clark and Sons in Cambridge, Massachusetts, had manufactured an $18\frac{1}{2}$ inch refractor which ended up at Dearborn Observatory of the (old) University of Chicago. In 1868 Thomas Cooke, an Englishman, constructed a 25 inch for R. S. Newall, a wealthy English amateur. Five years later, the Clarks completed the Naval Observatory 26 inch, which was followed in 1880 by the Vienna 27 inch, built by Howard Grubb. This in turn was followed by a $29\frac{1}{2}$ inch by Gautier of Paris.[30]

While the largest of the just-mentioned telescopes were being built, O. W. Struve was making plans of his own. He wrote to Newcomb in 1879:

> Last summer, when I suggested to the government that Pulkowa should be provided with a refractor which would be at least equal to the most powerful in existence, corresponding to the idea of the founding of the observatory, I particularly emphasized how much the optical power of the Washington refractor superseded that of our present telescope. Along these lines I expressed the desire to have an instrument with an aperture of 30–32 inches ... The government has granted me exactly

Table 5.2. *Directors of Pulkovo Observatory*[29]

Friedrich Georg Wilhelm Struve (1793–1864)	1839–1861
Otto Wilhelm Struve (1819–1905)	1862–1889
Fëdor Aleksandrovich Bredikhin (1831–1904)	1890–1895
Jöns Oskar Backlund (1846–1916)	1895–1916
Aristarkh Apollonovich Belopol'skiĭ (1854–1934)	1916–1919
Aleksandr Aleksandrovich Ivanov (1867–1939)	1919–1931
Anton Donatovich Drozd	1931–1933
Boris Petrovich Gerasimovich (1889–1937)	1933–1937
Sergeĭ Ivanovich Beliavskiĭ (1883–1953)	1937–1943
Grigoriĭ Nikolaevich Neuĭmin (1886–1946)	1944–1946
Nikolaĭ Nikiforovich Pavlov (1903–)	1947
Aleksandr Aleksandrovich Mikhaĭlov (1888–1983)	1947–1964
Vladimir Alekseevich Krat (1911–1983)	1966–1979
Kirill Nikolaevich Tavastsherna (1921–1982)	1979–1982
Vladislav Mikhaĭlovich Sobolev (1928–)	1982–1983
Viktor Kuz'mich Abalakin (1930–)	1983–

> 250 000 rubles (more than would be necessary to pay the manu-
> facturers), in order to be able to ensure against any accidents and to make
> sure that the auxiliary equipment is as good as possible, as well as for the
> construction of the dome and the upgrading of the building of our old
> refractor at the same time.[31]

It was eventually decided that the Clarks would manufacture an objective of 30 inches. Repsold would provide the mounting. Struve visited the United States in 1879 and 1883 to oversee the venture, and the telescope was put into operation in 1885. To Mrs Newcomb, Struve wrote:

> [June 27, 1885] was a glorious day for Pulkowa. The Emperor
> [Alexander III] and the Empress came here with the Crown Prince [the
> future Nicholas II] and his brother to see the new refractor and admire
> the combined work of the Clarks and Repsolds. The state of the sky was
> not favourable. Nevertheless, the Imperial family spent here 3 hours and
> on leaving expressed their satisfaction by the promise to visit us again
> very soon.[32]

Once again Pulkovo had the largest telescope in the world, but not for long. The Clarks manufactured the objectives for the Lick Observatory 36 inch (1888) and the Yerkes Observatory 40 inch (1897), which we will discuss in Chapter 6.

At this time Pulkovo also became involved with astrophysics – the investigation of the physical nature and evolution of celestial bodies by analysis of the light which reaches us. This would involve photography, photometry and spectroscopy, plus mathematical and physical models of the nature of matter under different conditions of composition, density, temperature and pressure. The first astrophysicist to work at Pulkovo was the Swede, Bernhard Hasselberg, who worked there from 1872 to 1889. He introduced photographic methods at Pulkovo, established an astrophysical laboratory, and conducted laboratory experiments on the luminescence of gases and the absorption spectra of chemical elements and compounds. He was able to demonstrate the existence of carbon compounds in comets from his laboratory experiments and observations at the telescope. However, astrophysics was a very young discipline; one might date its birth to 1859 when Kirchhoff and Bunsen unravelled the nature of the dark lines in the Sun's spectrum. Positional astronomy had a long and accomplished history, and an old observationalist like Otto Wilhelm Struve, whose primary love was the measurement of double stars, was conservative in his estimate of the state of astrophysics. In his review of the activity of the first 50 years of the Pulkovo Observatory he wrote:

> As far as astrophysics is concerned, the inclusion of it in the sphere of activity of the observatory was a necessity of the time, for without it our activity for practical astronomy would soon become incomplete. Also [astrophysics] in some of its parts is a completely new branch of investigation, and it is still far from that accuracy in observing and from conclusions based on this observing compared with precise astronomy, which represents almost rigorously mathematical motions of celestial bodies, but the rapid development of astrophysics in the last two decades and its successes in the mentioned direction [increased accuracy] clearly show its value for astronomy, and the successes serve as a guarantee of further improvement regarding accuracy.[33]

Yet more was afoot than the growth of another branch of work at Pulkovo.[34] For 50 years since its founding in 1839 it had been directed by the Struve family, father and son, who had acquired coworkers of Swedish or Baltic German descent. Would the observatory directorship pass on to the third generation of Struves? Would major changes occur which would greatly affect the value of the work, for better or for worse?

Throughout the 1880s there had been increasing friction between O. W. Struve and the Academy of Sciences. He had threatened to resign in 1887 but Tsar Alexander III requested that he stay on through the 50 year jubilee of the observatory, with the provision that he might collaborate on suggestions for the new director. O. W. Struve hoped that his son Hermann

Struve, who was qualified for the job, would succeed him. This was apparently the Tsar's wish as well, but Hermann was busy observing the moons of Saturn with the 30 inch refractor (the work for which he would win the RAS Gold Medal in 1903), and Hermann was unwilling to curtail his research to take on administrative duties. The Moscow astronomer Fëdor Bredikhin was appointed director in 1890, but it was clear that if Bredikhin some day no longer desired the job, Hermann would succeed to the directorship. But one can tell from Bredikhin's first director's report that times were changing:

> To the alumni of all Russian universities ... must be afforded, within the limits of possibility, free access [to the observatory] ... The recruitment abroad of scientists for its staff must and can be stopped forever.[35]

In 1895 there was an administrative crisis at Pulkovo. Concerning this O. W. Struve wrote to Newcomb:

> Bredikhin, who assumed the directorship five years ago only with aversion, because as a rich man he did not want the position and did not really take up the reins of command, declared last Christmas that he wished to leave in order to return to his estates. As his successor he suggested, *primo loco* and *ex aequo*, Backlund [whom O. W. Struve considered unqualified] and my son Hermann, but the administration of the Academy, upon whom the filling of the position unfortunately depends and which has been hostilely disposed to the Struves for a long time, determined from the outset that only the former would be considered in the balloting. The matter was thereby decided and Hermann has resolved to leave his position at Pulkovo, as it is a matter of honour for him not to work there if Backlund is director. However, the formalities of the vote took three months and gave Hermann some time to find another position ... We have nothing good to expect from the directorship of Backlund.[36]

Because of the biases that the Academy had shown in favor of those of Baltic German origin (e.g. when the famous chemist Mendeleev was passed over for a position at the Academy in 1880) by 1895, after the deaths of Alexander III and the President of the Academy of Sciences during the Mendeleev affair (Count Dmitriĭ Tolstoĭ), there were clamorings for reversals of favoritism shown to Baltic Germans. Though Backlund was not Russian, apparently his conflicts with O. W. Struve helped promote him to the directorship. Hermann Struve accepted a post in Königsberg and later became the director of the Berlin–Babelsberg Observatory. His brother Ludwig, also an astronomer, did find a job within the empire, becoming director of the KHarkov Observatory.

Ever since Russia's emergence as a progressive empire under Peter the Great, foreigners primarily of German origin had played important roles there. This included government, architecture, and the military, as well as science. To this day there are manifestations of an ideological struggle between two primary points of view. In times past the group known as the 'Westernizers' would advocate the adoption of the 'best' ideas regardless of origin and try to entice the best qualified to Russia, where they would figure prominently. Among such recipients of opportunities brought about by such thinking would be Daniel Bernoulli and Leonhard Euler, Swiss mathematicians who were invited to St Petersburg during the early years of the Academy of Sciences. The second group would be the 'Slavophiles', who believed in Russia's destiny to blaze its own trail without following the 'fashions' of the west. The third Pulkovo director, Bredikhin, quoted above, would fall into the latter category. Obviously, implementation of either scheme to the exclusion of the other would have serious repercussions for

Fig. 5.10. Pulkovo Observatory was completely destroyed during the siege of Leningrad in the Second World War. It was reconstructed basically according to the original plan of Briullov, except that hemispherical domes replaced the original octagonal turrets. The reconstruction was completed in 1954. This picture was taken by the author in 1974.

the scientific goals at an institution like Pulkovo Observatory. Whereas O. W. Struve avoided contact with Russian universities, had an outwardly hostile relationship with the Academy of Sciences, and hired mostly foreigners, Bredikhin's notion to hire Russians on the basis of nationality could turn out just as bad. Such notions are just as serious today with jointly owned telescopes being built and run throughout the world.

Because of the First World War, the Russian Revolution, the worldwide Depression, and the Second World War (during which Pulkovo Observatory was entirely destroyed during the siege of Leningrad – some of the instruments having been transferred early on and thus were saved), the history of the observatory in this century, described in detail elsewhere,[37] must be painted in different colors. By 1954 the observatory had been rebuilt according to the basic plan of Briullov (Fig. 5.10). It remains the Principal Astronomical Observatory of the Academy of Sciences of the USSR, continues its positional work, has undertaken solar astronomy and radio astronomy, and serves to train astronomers and guide the astronomical research of the Soviet Union.

6

Harvard, Lick and Yerkes and the rise of astrophysics

The present-day Bible of astronomy is the *Astrophysical Journal*, a journal which so much embodies the cutting edge of astronomical research that some astronomers would define the properly equipped astronomical library as having two copies of the '*Ap. J.*' and no other technical journals whatsoever! The driving force behind the creation of the *Ap. J.* was George Ellery Hale, a pioneering solar physicist and the greatest entrepreneurial observatory builder of modern times. The first meeting of the editorial board of the *Ap. J.* took place on 2 November 1894 in New York. Present were Hale, Henry Rowland, Albert Michelson, Charles Young, Charles Hastings, E. C. Pickering and James Keeler. The subtitle of the *Ap. J.* – *an International Review of Spectroscopy and Astronomical Physics* – and the accomplishments of these men let us distinguish quickly the new astronomy of that age from the astronomy of the previous centuries.[1,2] Whereas the old astronomy was concerned with counting the stars, producing catalogues of positions and producing ephemerides for the Sun, Moon and planets, the new astronomy was concerned with the chemical and physical composition of celestial bodies, and the application of laboratory methods from physics, chemistry and, today, even biology.

'Astrophysics' for the longest time meant almost exclusively 'optical spectroscopy'. In the early nineteenth century the optician Joseph Fraunhofer mapped the dark lines in the Sun's spectrum. It was not until 1859 that Gustav Kirchhoff and Robert Bunsen discovered the nature of these dark lines and founded spectral analysis. It was found that there were three kinds of spectra: (*a*) continuous spectra, which arose from a heated solid, liquid, or dense gas; (*b*) bright line, or emission spectra, which are produced by heated low-density gas; and (*c*) dark line, or absorption spectra, which are produced by a source of continuous radiation (as in *a*), with cooler gaseous material inbetween the continuum source and the observer – the absorption lines are indicative of the composition of the cooler gas. Kirchhoff and Bunsen found that each element is capable of emitting the radiation that it absorbs. The D lines of sodium, which appear in absorption in the yellow portion of the Sun's spectrum, appear as yellow emission lines in a candle flame that has salt sprinkled into it.

Not only did spectral analysis allow one to investigate the composition of stars, but one could also investigate the motions of stars along the line of sight. In the 1840s Christian Doppler and Hippolyte Fizeau elaborated the principle whereby an object receding from you should have spectral lines shifted toward longer wavelengths (the red shifts), while for objects approaching you the spectral lines are shifted toward shorter wavelengths (blue shifts). The measurement of these stellar radial velocities (from the astrophysicists) combined with proper motion data on stars (their angular motions across the line of sight, from the classical astronomers) allows investigations of the three-dimensional movements of stars in the galaxy. This is a good example of a marriage of the old and new astronomy.[3]

In the latter half of the nineteenth century, there were only a few astrophysicists in Europe, among them: the Englishman William Huggins, who discovered in 1864 that some unresolvable nebulae were clusters of stars (as they gave washed out absorption spectra), while others were gaseous (as they gave emission spectra); J. Norman Lockyer, another Englishman, and Jules Janssen, the founder of the Meudon Observatory, who independently discovered in 1868 how to observe solar prominences without having to rely on a total solar eclipse; J. K. F. Zöllner, who invented the visual photometer (used for directly measuring the brightness of stars) and the reversion spectroscope (which enhanced spectral line shifts, allowing more accurate radial velocities to be measured); Hermann Carl Vogel, a student of Zöllner, the director of the Potsdam Astrophysical Observatory from 1882 to 1907, who pioneered the measurement of stellar radial velocities and was the discoverer of some of the first known spectroscopic binary stars;* and the Italian, Angelo Secchi, who devised one of the first stellar spectral classification schemes.

In spite of this list of illustrious scientists, astrophysics has been a particularly American development. 'American money and technology, applied at fine observing sites in the favorable climate of California [and elsewhere too], enabled the United States to overtake Germany and Great Britain, and become the world leader in observational astronomy'.[4]

---

* The duplicity of such a pair of stars is inferred from the pairs of spectral lines that shift back and forth owing to Doppler's principle and the orbiting of the stars about their mutual center of mass.

Fig. 6.1. The Astrophysical Observatory of Potsdam. From Winterhalter, 1889, facing p. 259.

The first board of directors of the *Astrophysical Journal* constituted quite a powerful line-up of American pioneers of the new astronomy (lacking perhaps only Samuel P. Langley, who invented the bolometer, was director of the Allegheny Observatory, and in 1890 founded the Smithsonian Astrophysical Observatory). Henry Rowland, the inventor of the concave diffraction grating, and Charles Hastings both were faculty members of Johns Hopkins University (founded 1876), the first graduate research institution in the United States. Charles Young discovered the 'reversing layer' of the Sun's atmosphere in 1870 (the layer between the photosphere and the chromosphere in which the spectrum changes from an absorption to an emission spectrum). Albert Michelson was the first American recipient of a Nobel Prize in physics (1907). His ether-drift experiments led to Einstein's Theory of Relativity. He invented the interferometer and was the first to measure directly the diameters of stars. He determined a very accurate value of the speed of light. Michelson served on the faculty of the second American graduate research institution, the University of Chicago (founded 1892). Hale was the founder of the Yerkes and Mt Wilson Observatories (which we shall discuss in this chapter and the next). Keeler, who was a student of Hastings at Johns Hopkins, is considered the co-founder of the *Astrophysical Journal*. He became the second director of Lick Observatory. E. C. Pickering, like Simon Newcomb, was one of the deans of American astronomy at the turn of the twentieth century. Whereas the Naval Observatory embodied the developments of the old school of astronomy represented by Newcomb, Pickering's Harvard College Observatory, where he served as director from 1877 to 1919, embodied the new. The fact that the United States had two major institutions by the 1870s shows how far American science had come in a span of 50 years.

In 1825 President John Quincy Adams, in his first annual message to Congress, bemoaned the fact that there were 130 '*light-houses of the skies*' in Europe, while none was to be found in America.[5] This situation began to change in the 1830s as observatories sprouted up in many places. Yale University Observatory (founded 1830) had a 5 inch refractor by Dollond, which was used to observe the 1835/6 appearance of Halley's Comet. The existence of the United States Navy's Depot of Charts and Instruments (1830–42) led directly to the construction of the US Naval Observatory (1844), whose instrumentation included a 5.5 inch transit instrument, a 4 inch mural circle, a 5 inch prime vertical transit, and a 9.6 inch refractor by Merz and Mahler (the largest instrument there until the completion of the 26 inch Clark refractor in 1873).[6,7] The Hopkins Observatory of Williams College (Williamstown, Massachusetts), founded in 1836 (some sources say 1838), contained a $7\frac{1}{2}$ inch refractor by Alvan Clark and a $3\frac{3}{4}$ inch transit by Troughton & Simms. The Hudson Observatory of Western Reserve College, Cleveland, Ohio (1837); the Philadelphia High School Observatory (1839); the West Point Observatory (1839), which had a $9\frac{3}{4}$ inch refractor by Henry Fitz; the Georgetown College Observatory (1843); and the Cincinnati Observatory, built by public subscription in 1843, which contained an $11\frac{1}{4}$ inch refractor by Merz and Mahler – these were some of the observatories built at that time.[8]

Harvard, too, wanted an observatory and induced William Cranch Bond of Dorchester, Massachusetts, to become Astronomical Observer to the University. Bond's family business had been clock making, which afforded him a comfortable lifestyle, but from 1823 to 1839 he spent increasing amounts of time on astronomical pursuits carried out from his own private observatory. He used meridian instruments and obtained contracts for the rating of chronometers such as those for the United States Exploring Expedition of 1838–42. He clearly had a talent for building and using astronomical instruments, and we must commend his love of, and commitment to, astronomy, for his new post at Harvard carried with it time consuming responsibilities but no salary. Bond brought his own instruments to

Cambridge, where they were set up at Dana House along with a few instruments owned by Harvard. While Dana House was fitted with a revolving dome on top, it was expected that some observations would be made out of regular windows, much like the inadequate plan of the original Paris Observatory. A better observatory was clearly needed.

In the wake of public interest over the great comet of 1843, one David Sears offered $5000 for the building of an observatory tower provided that an additional $20 000 could be raised. In the span of 6 weeks, 82 individuals, 7 firms and 3 non-profit institutions donated the required total. Sears soon donated almost $10 000 more to serve as the nucleus of a capital fund for the permanent endowment of the observatory.[9]

Given the established reputation of the 9.6 inch refractor at Dorpat, which had been made by Fraunhofer, and the (then) recent establishment of the 'astronomical capital of the world' at Pulkovo, whose 15 inch refractor by Merz and Mahler (the successors of Fraunhofer), was the largest and considered the best telescope in the world, Harvard decided to turn to the same firm and 'procure an instrument fully equal to the best they have ever made, and even superior, if that is possible'.[10]

The new observatory on Garden Street, then at the outskirts of Cambridge, was completed in 1846. 'From that year, in fact, one may date the effective operation of the present Harvard Observatory'.[11] At that time, by means of another public subscription, W. C. Bond was finally given a salary, $1500 a year for 2 years, while $640 was provided for the salary of an assistant. George Phillips Bond, the third son of William Cranch Bond, was appointed as assistant. In the tradition of great astronomical family dynasties – the Cassinis, Herschels, and Struves – we must also include the two Bonds. But while the other astronomical dynasties just mentioned were for the most part sequentially ordered in time, 'the scientific collaboration [between the two Bonds] began so early that it is often difficult to separate their contributions'.[12]

The 15 inch 'Pulkovo twin' achieved first light in the summer of 1847.[13] One of the first principal objects of study that year and in subsequent years was the Orion Nebula, which William Cranch Bond claimed to have resolved into stars. In fact, he did observe previously unknown faint stars, but they were embedded in nebulosity, allowing him to think that he had indeed resolved the object into stars, much like observing a globular cluster with a medium-sized telescope. But W. C. Bond was mistaken, leaving himself and the reputation of Harvard College Observatory open to criticism. Otto Struve at Pulkovo was one such critic, and G. P. Bond's extensive work on the Orion

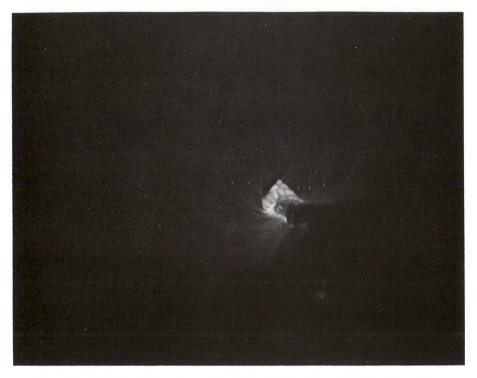

Fig. 6.2. G. P. Bond's drawing of the Orion Nebula. From Holden, Edward S., *Monograph of the Central Parts of the Nebula of Orion, Appendix I to Astronomical and meteorological observations made during the year 1878, at the United States Naval Observatory*, Washington, DC: Government Printing Office, 1882; original from Harvard *Annals*, 5, 1867. Photograph courtesy of the Lick Observatory.

Nebula, in addition to having its own intrinsic motivation, was carried out to try to vindicate the previous work done at Harvard and the family reputation.[14]

The first major discovery made with the 15 inch refractor was that of the eighth satellite of Saturn, Hyperion, discovered on 16 September 1848. It is to be regarded as a joint discovery by the two Bonds. Their interest in Saturn and the systematic observation of that planet led two years later to George Bond's discovery of the Crêpe Ring, independently discovered in England by William R. Dawes and by William Lassell.

These two discoveries made European astronomers sit up and take notice that colleagues/competitors in astronomy now were active on the other side of the Atlantic. From the standpoint of the later astrophysical methods, we must rate as more significant the early experiments with photography in America. Following in the footsteps of his European counterparts Daguerre,

Niepce and W. H. Fox Talbot, already in 1840 the American Dr John W. Draper had obtained a daguerreotype of the Moon. Given the extreme commercial success of photography, photographic studios were soon thriving in many cities in Europe and America. In Boston the firm of John Adams Whipple and J. Wallace Black did excellent work, which could not help but be noticed by the nearby Harvard astronomers. Shortly after the installation of the 15 inch refractor, Whipple and the Bonds began experiments with photographing the Moon and Sun, and obtained the first photograph of a (nighttime) star, Vega, on 16 July 1850. They found that they could photograph first magnitude stars with exposures of 1 or 2 minutes, while they were unable to obtain images of second magnitude stars. There were obvious problems with guiding, as any amateur will testify if trying to take an exposure longer than a few seconds.

How does one take a daguerreotype? It is a silver plate, which is polished and coated with silver iodide by submitting it to iodine vapor for about 20 minutes. The image is projected onto the plate with a camera, the image is developed with mercury vapor, and the plate is fixed by means of immersing it in a solution of sodium thiosulphate. This was the principal means of obtaining photographs at Harvard until 1857, when Whipple and Black began taking lunar photographs with the wet-plate collodion process. This involves spreading a viscous solution of collodion (pyroxylin dissolved in ether or alcohol), laced with a soluble iodide, onto a glass plate. The plate is exposed while wet, developed in a solution of pyrogallol containing acetic acid, and fixed with a concentrated solution of sodium thiosulphate. While this allowed fainter objects to be photographed (the plates were more sensitive), the exposures were limited to the drying time of the collodion solution.[15]

In 1857, in a review of the photographic experiments at the Harvard College Observatory, G. P. Bond wrote:

> It is reasonable to suppose that on some lofty mountain and in a purer atmosphere we might, with the same telescope, include the eighth magnitude ... To increase the size of the telescope threefold in aperture ... would increase the brightness of the stellar images, say eightfold, and we should be able to photograph all the stars to the tenth and eleventh magnitude inclusive. There is nothing, then, so extravagant in predicting a future application of photography to stellar astronomy on a most magnificent scale.[16]

During the 10 years of their experiments they obtained some 70 daguerreotypes, and more than 200 wet-collodion plates of the Moon, stars, and planets. Very little photographic work was done at Harvard inbetween

1860 and 1880. Dry gelatino-bromide plates were invented in 1874 by W. Abney. Huggins was one of the first to use them for astronomical purposes, and it was clear that the new dry plates would allow projects to be carried out 'on a most magnificent scale'. Not only were the plates more sensitive, but one could now take exposures several hours long, if only one could guide accurately.[17] At Harvard Pickering later planned extensive programs for sky photography and for stellar spectroscopy.

William Cranch Bond served as director of Harvard College Observatory from 1839 to 1859. His son, George Phillips Bond, succeeded him in this position, but died in 1865 while in his fortieth year. During their tenure they accomplished a great deal, but one of their continuing problems was that of funding. A rough gauge of their financial situation is based on the fact that their annual income, from interest drawn on contributions to the Observatory, was about $6000 a year in 1860, from which salaries, equipment, publication fees, and costs of possible expeditions would be paid. By contrast, the national facilities had much more. At the Royal Greenwich Observatory the Astronomer Royal was paid $5000 a year, and an additional $7000 per year paid the salaries of seven assistants. At the US Naval Observatory the salaries amounted to $35 000 annually. At Pulkovo Observatory the director earned $6700, four assistants received $10 000, and the expeditions, publications, and new equipment were also amply provided for.[18]

The third director of the Harvard College Observatory (1866 to 1875) was Joseph Winlock, who is primarily remembered for improvements to the instrumentation of the observatory. He ordered a new meridian circle from Troughton & Simms, which was delivered in 1870; he ordered a portable transit instrument to be built in Europe under the direction of Otto Struve; in 1867 he initiated a time service, providing time signals to the railroads, which gave the observatory much-needed income.[19] Under his direction many thousands of double star measurements were made with the 15 inch refractor. He collaborated with the *Astronomische Gesellschaft* in the production of improved data for catalogues of stellar positions. Research at Harvard was not just along the lines of positional work. Winlock resumed experiments in photography and instituted new ones in spectroscopy. Samuel P. Langley, one of the pioneers of astrophysics, got his start under Winlock at Harvard. Winlock also headed up two solar eclipse expeditions. It was during the 1870 eclipse, observed by his party in Spain, that Charles A. Young discovered the flash spectrum and the solar reversing layer.[20]

The reader will recall that the Royal Greenwich Observatory, established in 1675, only came of age about 1750. Harvard College Observatory took 40

Fig. 6.3. Harvard College Observatory at the end of the nineteenth century.
Photograph courtesy of the Harvard College Observatory.

years to become a mature, well-established research institution. This gathering of momentum was due in large part to a tremendous increase in its funding, which naturally provided for an enlargement of its activity, and it was also due to the personality and administrative talent of its fourth director, Edward C. Pickering, who became Director and Phillips Professor of Astronomy in 1877, under whose direction Harvard embarked on extensive programs of visual photometry of stars, stellar spectroscopy, and stellar photography. So much work was carried out that we can only provide a rough outline here.[21]

After 2 years of experimenting with variations of the Zöllner photometer, Pickering and his colleagues developed a meridian photometer which allowed their chosen comparison star, Polaris, to be measured alongside any star on the meridian by means of a series of mirrors and prisms. They adopted Pogson's definition of magnitudes, whereby 5 magnitudes corresponds to a factor of 100 in relative brightness (hence 1 magnitude = $\sqrt[5]{100} \approx 2.512$ in relative brightness). Apparent magnitude 2.1 was assigned to Polaris. Though it was later discovered that Polaris is a small amplitude variable star (varying 0.14 magnitude), this did not negate the value of Harvard's visual photometry, which was the first such comprehensive effort carried out over the whole sky. From 1879 to 1882, 4000 stars of sixth magnitude and brighter were measured. The results were published in 1884 in volume 14 of

**Fig. 6.4. Edward C. Pickering (1846–1919), fourth director of Harvard College
Observatory. Photograph courtesy of the Harvard College Observatory.**

the *Annals* of the Harvard College Observatory. As soon as this project was
finished, Pickering began another with an improved photometer able to
measure stars as faint as seventh magnitude. When all the stars visible from
Cambridge were measured, this instrument was sent to Peru for measurements of the southern stars. Volumes 50 and 54 (1908) of the Harvard
Annals presented the *Revised Harvard Photometry*, containing the magnitudes
of 45 000 stars. Pickering had been very active in this project, personally
making 1.5 million photometric estimates.

From 1886 to 1889 the Harvard College Observatory (HCO) obtained four
new principal sources of funds. The Robert Treat Paine Fund of $300 000
provided for the director's salary and observatory expenses. (The Director
was now the Paine Professor of Astronomy at Harvard, while the assistant
director was the Phillips Professor.) The Henry Draper Fund also became
available in 1886. It was a memorial to one of the founders of astronomical
photography, Henry Draper (the son of John W. Draper), who had obtained

the first photograph of a nebula (the Orion Nebula) in 1880.[22] Draper died at the age of 45 in 1882. His widow, Anna Palmer Draper, had been his principal assistant (much like the team of Sir William and Lady Huggins in England); upon Draper's death, Mrs Draper sought a means of having his work carried on. During her lifetime she regularly donated money to the Harvard College Observatory and, along with provisions in her will, the total was on the order of $400 000.

A third source of funds was from one Uriah A. Boyden, who donated $238 000 for a high altitude observatory, 'at such an elevation as to be free, so far as practicable, from the impediments to accurate observations which occur in the observatories now existing, owing to atmospheric influences'.[23] Finally, Miss Catherine Wolf Bruce, who donated a total of $175 000 to American and European astronomers during her lifetime, provided Harvard with $50 000 in 1889 for a 24 inch photographic refractor.

Rather than observing the stellar spectra one at a time, visually at the telescope, the Harvard plan was to place a large prism in front of a telescope's objective lens and photograph all the spectra of the stars in the field of view. These are *objective prism spectra*, which do not have much detail in them (being typically a few millimeters long on the plate), but obtaining the data this way allows one to record a great many spectra at a time while leaving the classfication of the spectra to daytime or cloudy-night hours. Volume 27 of the Harvard *Annals* (1890) contains *The Draper Catalogue of Stellar Spectra*, containing the spectral classes and magnitudes of more than 10 000 stars down to about eighth magnitude and north of declination $-24°$. The spectra were obtained with an 8 inch photographic refractor, the Bache telescope, the funds for which ($2000) were provided in 1885 by the National Academy of Sciences. The 1890 catalogue was primarily the work of Williamina Paton Fleming, who classified the spectra on the basis of the strengths of the hydrogen and calcium absorption lines (spectral types A through O, with J omitted, and including type Q for peculiar spectra).

A second spectral project being carried out at the same time involved the investigation of more detailed spectra of brighter northern stars. The data were taken with the 11 inch Draper photographic telescope, which was moved from Mrs Draper's estate to Cambridge in 1886. Much higher resolution was obtained by the use of up to four prisms in front of the objective, giving spectra a few inches long instead of a few millimeters. *The Spectra of Bright Stars*, published in 1897 (Harvard *Annals*, volume 28, Part I), contained detailed information on 681 bright stars. This was the work of Antonia C. Maury, who was Henry Draper's niece. She had decided that Fleming's spectral classification scheme was inadequate and devised one of her own. The class names were Roman numerals I to XXII, with three subclasses (a, b, and c) to denote a sequence of decreasing widths of the spectral lines. It turned out that classes I to XXI represented a temperature sequence of the stars, and that stars of subclass c were giant stars. Volume 28, Part II of the *Annals* (1901) gave the classifications (on Cannon's system – see below) for 1122 bright stars in the southern sky; this was work by Annie J. Cannon, based on plates taken in Peru.[24]

The third part of the spectral classification work was begun by Annie Cannon in 1911. Within 4 years, from spectra taken in Cambridge and at Harvard's station in Peru, she classified the spectra of all the stars brighter than magnitude 10.5 – 225 300 stars in all! These results were published in volumes 91–9 of the Harvard *Annals* (1918–24) – the famous *Henry Draper Catalogue*. Cannon's *Henry Draper Extension* (1925–36), published in volume 100 of the *Annals*, contained spectral classifications of 47 000 fainter stars. The classification scheme was a combination of Fleming's and Maury's. Cannon rearranged the alphabetic groups according to the temperatures of the stars, eliminated some letters by combining some groups, and added types R (in 1908) and S (in 1922), giving the well known sequence O, B, A, F, G, K, M, R, N, S.* Cannon also added decimal subgroups (e.g., G2 or A5). The Harvard classification scheme was adopted by the International Union for Cooperation in Solar Research in 1910. It was adopted by the International Astronomical Union in 1922.

In response to an aborted attempt to establish a Harvard station at Wilson's Peak near Pasadena, California in the year 1889–90, Pickering chose South America for the fulfillment of the goal of the Boyden Fund. Primarily under the supervision of Solon I. Bailey (1889–91, 1893–7, and 1899–1904), Harvard established an extremely productive observing station at Arequipa, Peru (elevation 8055 feet), which continued in operation until 1918 when the First World War forced it to be shut down. In 1927 the equipment was moved to Bloemfontein, South Africa.[25]

The work carried out in Peru made possible by the Draper, Boyden, and Bruce bequests turned the Arequipa–Cambridge axis into a great astronomical data pipeline or a permanent harvest being sent north to the processing plant. The photometric and spectroscopic projects have already been mentioned. In the area of astronomical photography, one of the many firsts for Harvard was the first photographic map of the entire sky, made in Cambridge and Arequipa with 2.5 inch lenses. Each plate was 30° on a side

---

* This is usually remembered by the mnemonic Oh, Be A Fine Girl, Kiss Me Right Now Sweetheart, but I prefer Oh, Bring A Full Grown Kangaroo, My Recipe Needs Some, or Oh, Beastly And Ferocious Gorilla, Kill My Roommate Next Saturday. (See *Mercury*, November–December 1976, p. 13, and July–August 1977, p. 21.)

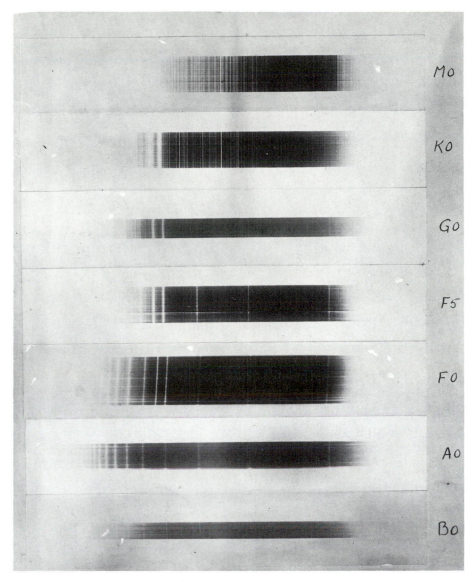

Fig. 6.5. Harvard spectral classes by Annie Cannon. From top to bottom stars increase in temperature. Photograph courtesy of the Harvard College Observatory.

and showed stars down to twelfth magnitude. This was published in 1903.

The Bache 8 inch, used in Peru since Bailey's first trip there in 1889, was routinely used to photograph large areas of the sky, resulting in a major early part of the Harvard plate collection. This now amounts to 300 000 plates, a history of the night sky duplicated nowhere else and particularly useful for *ad hoc* studies of unusual variable stars.

Fig. 6.6. The Boyden station of the Harvard College Observatory, situated at Arequipa, Peru. In 1927 the equipment was moved to South Africa. Photograph courtesy of the Harvard College Observatory.

The Bruce 24 inch refractor, first set up in Peru in 1896, produced 14 by 17 inch plates covering 25 square degrees. Pickering had planned on photographing the entire sky with this instrument instead of collaborating in the *Carte du Ciel* project (Chapter 4), a decision which brought forth contemptuous remarks in the pages of *The Observatory* in 1889. One would at least expect the mapping to proceed more rapidly, since each plate with the Bruce telescope covered six times the area of the *Carte du Ciel* plates. As for the quality of the results, Pickering had every reason to expect his telescope to perform well, as the objective was made by Alvan Graham Clark. Pickering felt very strongly that the *Carte du Ciel* project was organized all wrong; regarding criticism that Pickering did not know what he was doing, he replied. 'I do not feel that I am to blame if with one telescope we do more and better work than with seventeen of the usual form'. Furthermore, he stated to Catherine Wolf Bruce, 'It is the first time I have ever felt obliged to apologize on account of the expected *excellence* of a piece of work'.[26]

A complete map of the southern sky was produced. The faintest stars visible on the plates were about 14th magnitude. The Bruce telescope was a success.

One of the outstanding discoveries made with photographs taken with the Bruce telescope was a discovery by W. H. Pickering (the brother of E. C.

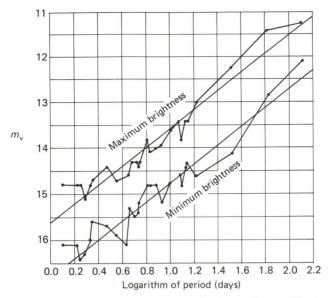

m_v

Logarithm of period (days)

Fig. 6.7. Period–luminosity diagram for Cepheids in the Small Magellanic Cloud. Henrietta Leavitt discovered that Cepheid variables with longer periods are brighter. Shapley calibrated the absolute magnitudes of Cepheids and, based on the distances to Cepheids in globular clusters, proposed a radically different model of the structure of our Galaxy. After Herrmann, Dieter B., *The History of astronomy from Herschel to Hertzsprung*, Cambridge University Press, 1984, Fig. 44, p. 140; original from *Harvard College Observatory Circular*, No. 173, 1–3, 1912.

Pickering). In April of 1899 he identified the ninth moon of Saturn (Phoebe) on plates taken 7 months previously. Thus Harvard could be proud that its astronomers had discovered the eighth and ninth moons of Saturn.

The greatest result derived from plates taken with the Bruce telescope arose from photographs of the Magellanic Clouds. At Harvard, Henrietta Leavitt pored over the plates and in 1908 was able to publish a list of 1777 variable stars in the two Clouds. Most of them were Cepheid variables, named after the prototype, δ Cephei. These stars vary in brightness by about one magnitude, with periods inbetween about 1 day and 100 days (typically 5–15 days). From plates taken of the Small Magellanic Cloud, Leavitt discovered that the Cepheids with the longer periods were predominantly brighter. Since all the stars in a distant cluster are approximately at the same distance, this implied an underlying physical relationship between the intrinsic brightness (absolute magnitudes) of the stars and the periods of variation.[27] It is no understatement that this serendipitous discovery was the most important observational finding of the turn of the century, for it allowed the calibration of the distances to clusters of stars throughout the Galaxy, radically changing our conception of its size and shape.

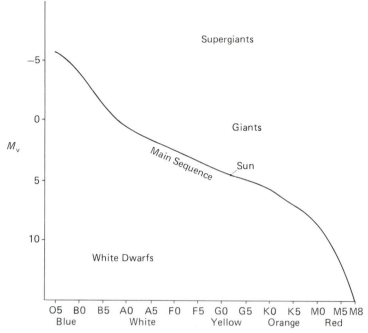

Fig. 6.8. The Hertzsprung–Russell diagram, plotting the absolute magnitudes of stars versus their spectral types. Luminosity increases going up in the diagram, while temperature increases to the left. Most stars are to be found on the Main Sequence. It is easy to understand that the stars are of different sizes. Consider a G dwarf (Main Sequence star) with $M_v = +5$ and a G giant with $M_v = 0$. Since the spectral type is indicative of temperature, and the temperature dictates how much light is given off per unit area, each square centimeter of the G giant gives off the same amount of light as a square centimeter of the G dwarf. The G giant is 5 magnitudes brighter, which corresponds to a factor of 100 times more luminosity. Therefore, it must have 100 times more square centimeters on its surface. A ratio of 100 in the surface areas of the stars means that the giant star has a radius 10 times that of the dwarf star. A supergiant G star of $M_v = -5$ would be 10 times larger still.

The discovery of variable stars was a particularly Cantabrigian affair as well. Henrietta Leavitt alone discovered about 2400 of them. In 1939 Harvard's 10 000th variable star was discovered by Annie Cannon. E. C. Pickering played a prominent role in the establishment of the American Association of Variable Star Observers in 1911, which still organizes the worldwide observation of variable stars of all kinds, primarily by means of visual estimates of star brightness from charts calibrated with photographic plates.

Another extremely significant finding of the early twentieth century was the discovery by Hertzsprung and Russell of the relationship between the absolute magnitudes and temperatures of the stars (Fig. 6.8). This was crucial for studies of stellar evolution, and was made possible by stellar parallaxes and double star measurements from the old (positional) astronomy combined with the spectral types of stars from the new astronomy of astrophysics, in particular Harvard spectral studies.

In summarizing the Pickering years at Harvard, one must conclude that the large increase of the observatory's endowment and the dynamic leadership of E. C. Pickering brought about an immense, and immensely important, data base for the study of the stars. The study of stellar spectra, the photography of the heavens, and the photometry of stars are all fundamental for the new astronomy based on astrophysics. Pickering's genius lay in being able to mobilize his troops to carry out expeditions overseas, to endure the near-sweatshop atmosphere of data reduction then required, and to encourage the involvement of amateurs to the limit of their energies in the monitoring of variable stars. Pickering was one of those scientific diplomats of the turn of the century who could successfully appeal to wealthy benefactors; his ability to be both benevolent patriarch of the observatory and great organizer of choice astronomical projects of the day must make us look on his reign as one of the great kingships of astronomy.

The fifth director of HCO (1921–52) was Harlow Shapley, who was, from a scientific and administrative standpoint, one of the most prominent American astronomers of the first half of the twentieth century.[28] Shortly after finishing his doctorate at Princeton (1913) Shapley accepted a job at the Mt Wilson Observatory. Stopping at Harvard on his way to California, he discussed his research plans with Solon I. Bailey, who suggested that he make observations of variable stars in globular clusters. Shapley took up Bailey's suggestion. Whereas Henrietta Leavitt had discovered that Cepheids of differing periods in the Small Magellanic Cloud had certain *relative* brightnesses, Shapley's observations of nearby Cepheids and his observations of cluster Cepheids allowed him to determine the *absolute* brightnesses of these stars (i.e. he 'fixed the zero point' of the period–luminosity law). He then determined the distances to the globular clusters.

It was well known that the globular clusters were primarily distributed in one half of the sky, in the constellation Sagittarius in particular. Shapley assumed that the center of the globular cluster system coincided with the center of our galaxy, implying that we were viewing the galaxy from a position far removed from its center. On the basis of the distances to the globular clusters, derived by means of the period–luminosity law, he obtained a distance to the center of the galaxy of about 50 000 light years (about twice the modern value). The accuracy of his result did not matter as much as the qualitative result. For the first time in 100 years there was a major breakthrough in galactic astronomy. The results of William Herschel and, more recently, those of J. C. Kapteyn, had placed us near the center of a flattened ellipsoidal system of stars no larger than 30 000 light years in size. Shapley's qualitative results, which were relatively rapidly accepted by other astronomers, placed us far from the center of a comparatively giant star system nearly 300 000 light years in diameter.

I say that Shapley's *qualitative* results were accepted, because the quantative results were questioned for some time. This led to the twentieth century's Great Debate on the 'scale of the universe',[29] which took place in 1920 between Shapley and H. D. Curtis, a Lick Observatory astronomer.

Actually, calling it a debate is somewhat of a misnomer, since it was not a formal debate on a single topic. Shapley took the 'scale of the universe' to mean the scale of our galaxy. He discussed the distances to the globular clusters and his model of the Big Galaxy. Curtis took 'scale of the universe' to mean everything we can see, including what he considered to be extragalactic nebulae, the so-called 'island universes'. Curtis claimed that Shapley's globular cluster distances were wrong, and that our galaxy was not nearly as big as Shapley thought.

Shapley was reluctant to claim that the faint nebulae were separate galaxies (as we now call them), since proper motions supposedly measured by Adriaan van Maanen implied that they would be rotating at enormous speeds if at great distances. Curtis countered that van Maanen's data must be wrong, because observations of novae in the faint nebulae did place them at enormous distances. Furthermore, the derived linear sizes of the faint nebulae implied that they were all about the same size and about one-tenth the size of Shapley's Big Galaxy. It turned out that van Maanen's proper motions were wrong, Shapley's model of the galaxy was essentially correct (some of his quantitative errors had magically cancelled each other), and Curtis' idea of the existence of separate galaxies was correct.

There are astronomers who would claim that administrative work is the death of one's research career, and point to Shapley as a prime example. In his 7 years at Mt Wilson he had published over 100 papers, and as director of HCO his time was not so free to pursue research. But the 1920s and 1930s are referred to by Bart Bok as Shapley's 'glory years', during which time he discovered the first two dwarf galaxies, in the constellations Sculptor and Fornax, which are part of the Local Group of galaxies, and he published the Shapley–Ames catalogue of 1246 galaxies (1932), which included most galaxies brighter than photographic magnitude 13.

In 1923 a most significant change took place at Harvard – the inauguration of formal graduate study in astronomy. This lack had been the observatory's major weakness ever since its founding. While the director and his assistant were the Paine and Phillips Professors of Astronomy, no formal teaching had been done. From 1930 to 1945 approximately one-third of all PhDs earned in the United States were from Harvard.[30]

The first person to obtain a PhD for work done at Harvard was Cecilia Payne (later Payne-Gaposchkin). Her thesis, entitled 'Stellar Atmospheres', was published as *Harvard Observatory Monograph*, No. 1, in 1925. According to the American astronomer Otto Struve, 'It is undoubtedly the most brilliant PhD thesis ever written in astronomy'.[31] Based on the ionization theory pioneered by Meghnad Saha and the extensive spectral library at Harvard, Payne tied together theoretical ideas on spectral line formation with observational results. Since the end of the nineteenth century it was understood that the spectra of hot stars had helium and hydrogen lines, the cooler stars had spectra with many metallic lines, while the coolest stars had bands in their spectra. It was work such as Payne's that showed that all stars have remarkably similar compositions. It is primarily the temperature that gives rise to the observed spectrum. One of Payne's principal findings was that stars were primarily hydrogen and helium, but because the influential Princeton astronomer Henry Norris Russell (Shapley's thesis advisor) did not believe this result, Payne intentionally played it down and even stated that something must be wrong with her analysis. This is just one example in the history of science of an important new result being rejected solely on the basis of the authority of a critic.[32]

It is self-evident to the reader that women played an important role in making Harvard one of the principal astronomical capitals. While there were a number of women computers at Harvard and at Greenwich[33] in the late nineteenth century, we remember the work of Williamina Fleming, Antonia Maury, Annie Cannon, Henrietta Leavitt, and Cecilia Payne-Gaposchkin as being of primary significance on its own merits, independent of the guidance of such directors as Pickering and Shapley. While these women were given the 'privilege' of making contributions to astronomy at Harvard, they were not afforded the salaries or rank of their male counterparts. Though Cecilia Payne-Gaposchkin became the first female full professor at Harvard in 1956 and also the first department chairman, this step was long overdue.

While Shapley was director, the number of staff at the observatory tripled, as well as the number of active telescopes. In 1931 HCO established their Oak Ridge Station near the town of Harvard, Massachusetts. (Since 1952 it has been called the Agassiz Station.) A 61 inch telescope went into operation there in 1932. They established other stations in Colorado, New Mexico, and ran the Boyden station in South Africa. In addition to it being a center for galactic astronomy and theoretical astrophysics, Harvard also became involved with solar astronomy, meteor patrols and radio astronomy. The largest radio telescope operated at the Agassiz Station has been an 84 foot diameter steerable dish.

In 1955, while Donald H. Menzel was director of Harvard College Observatory, the Smithsonian Astrophysical Observatory moved from Washington, DC to Cambridge. This came about with Menzel's active participation. In 1973 the Harvard–Smithsonian Center for Astrophysics was created, which has over 200 scientists working on all the areas of astronomy so far mentioned, plus some new ones. The Center for Astrophysics was the headquarters for X-ray astronomy being carried out with the High Energy Astronomy Observatories. Cambridge astronomers are also involved in balloon-borne astronomy from a station in Palestine, Texas. Jointly with the University of Arizona, they operate the Whipple Observatory at Mt Hopkins, Arizona, the site of the Multiple Mirror Telescope (see p. 266).

I must now backtrack a century to pick up the thread of the development of astronomy at Lick Observatory, one of the other perennial centers of research.

Lick Observatory owes its origin to the generous miser James Lick (1796–1876), an eccentric man who made his fortune in the piano business in South America and in real estate speculation during the California gold rush days.[34] As he got on in years, Lick (who had no legitimate heir) sought a means by which to perpetuate his name. One of his early ideas was to build a pyramid larger than any in Egypt.[35] About 1860 the idea of an observatory began to percolate in Lick's brain. The first public announcement regarding Lick's intention to build an observatory was made on 20 October 1873 at a meeting of the California Academy of Sciences. Two days later Lick wrote to Joseph Henry, the director of the Smithsonian Institution, who had encouraged Lick to consider funding something scientific:

> . . . I have in contemplation the erection of an Observatory on this Coast which shall rank first of any in the World . . . [containing] a *Telescope* that shall be as near a perfect instrument as the best Scientific Knowledge and skill and human workmanship can make, and that shall surpass in power anything yet attempted.[36]

Lick first intended to place the telescope in downtown San Francisco,[37] but he was persuaded to consider a more rural, and more astronomically auspicious, site. For a time Lick was considering the shores of Lake Tahoe on the California–Nevada border.[38]

The first of three sets of Lick Trustees was appointed in the summer of 1874. The Lick Trust was charged with disposing of the old man's fortune of $3 000 000, of which $700 000 was set aside for the future observatory. When the observatory was completed, it was to be turned over to the University of Calfornia and be known as the Lick Astronomy Department of the University of California.

James Lick himself decided that the observatory was to be built at the 4200 foot summit of Mt Hamilton overlooking San José, California, providing that the county of Santa Clara build a 'first class' road from the Santa Clara Valley to the top of the mountain. The county was more than happy to oblige and did build the road. As the equipment would be hauled to the top by mule train, this stipulated the maximum grade of the road, which is the reason there are so many hairpin turns on it and why travel on the Mt Hamilton road can be an emetic experience one will remember fondly.

In December of 1873 Joseph Henry had written to Lick to point out that if the new observatory were devoted solely to positional astronomy, a refracting telescope would be advisable, but if the new observatory were to be involved with spectroscopy and 'the observation of the physical phenomena of the heavens and the earth', that a large reflecting telescope would also be necessary. Henry also suggested that the 'best talents of the world' be found and that Lick should 'at once secure a Director, of the best talents and practical skill in the line of astronomy that can be obtained'.[39] This sentiment was shared by Simon Newcomb of the US Naval Observatory, who had been called on for his advice in the matter of the new observatory. Newcomb suggested that the director 'should be chosen in advance of beginning active work, so that everything should be done under his supervision'.[40] Joseph Henry had suggested that J. N. Lockyer, the English astrophysicist, be nominated as director, clearly indicating that the work of the new observatory should be oriented along the lines of astrophysics. However, Newcomb nominated his protégé, Edward S. Holden, for the post, and after 1874 Holden was regarded as the likely eventual director.

The first site testing at Mt Hamilton was carried out in 1879 by S. W. Burnham, a double star observer from Chicago. Luckily for everyone the site turned out to be excellent. Indeed, it turns out that many of the world's major observatories are placed on mountaintops within 50 miles of the ocean. This often leads to very stable observing conditions for a sizeable fraction of the year.

It was decided that the telescope was to be a 36 inch refractor, the objective lens of which was to be made by Alvan Clark and Sons, the makers of the Naval Observatory 26 inch. Work steadily increased at Mt Hamilton starting in the summer of 1880.

The chairman of the third set of Lick Trustees was Captain Richard S. Floyd. The construction of the Lick Observatory was the result of his efforts more than anyone else. Newcomb in Washington, DC, was still the principal scientific advisor on the project. Holden had taken up the directorship of the Washburn Observatory of the University of Wisconsin but wrote to Floyd that he considered that position temporary. Holden was just biding his time to assume the role of Lick director. In 1885 the regents of the University of California offered Holden the job of president of the university with the understanding that he would resign that position and become Lick director as soon as the observatory was completed. Holden moved to California, where he could be close to the scene of action of the construction of the observatory.

Floyd, however, kept Holden off Mt Hamilton, since it was explicitly understood that the University of California would not have a direct hand in the observatory until it was actually handed over to the regents. By 1886 Captain Floyd was living at the mountain, and the construction of the observatory building was taking place in earnest. He had hired James E. Keeler, a young astrophysicist who had worked under Langley at the Allegheny Observatory, to serve as astronomer-on-site. Keeler acted as go-between for Holden and Floyd.

The observatory was duly completed and turned over to the regents of the University of California on 1 June 1888. Holden resigned as university president and moved to Mt Hamilton to become director. As his four assistants he retained Keeler and hired S. W. Burnham, John M. Schaeberle (a meridian circle observer), and Edward E. Barnard (a highly talented and maniacally enthusiastic visual observer and experienced photographer). In addition to the world's largest refractor, the new observatory had a 12 inch Clark refractor, a 4 inch combined transit and zenith telescope, a horizontal photoheliograph, and a 6 inch Repsold meridian circle. These smaller instruments had come into operation during the years 1881 to 1884. A time service, organized by Keeler and providing time signals to the Southern Pacific railroad, went into operation in 1887.

Lick Observatory opened its domes to the universe, having first-class equipment at an excellent site, with highly qualified personnel, and a director 'of the best talents and practical skill ... that can be obtained'. So

Fig. 6.9. Lick Observatory, not long after its founding. The stormy clouds are reminiscent of the general mood of the staff during the 10-year reign of the first director, Edward S. Holden. Lick was the first permanent mountaintop observatory, and thus a new experiment in coping with the isolation resulting from the pursuit of better seeing conditions. Photograph courtesy of the Mary Lea Shane Archives of the Lick Observatory.

naturally they got right to work and everything went well, right? Well, not exactly. Whereas Holden had been a very conscientious and energetic advisor for 14 years and had assembled an excellent staff, his own research orientation was not solid, and his personal manner eventually alienated everyone.[41] Just before assuming the directorship, Holden presented a bill for $6000 to the Lick Trustees for 'services rendered' since 1876; while expenditure of time and effort, and for travel, are routinely compensated today under the agreement that this is fair, Holden had no such agreement when he acted as advisor to the Lick Trust, and this act by Holden gave many the permanent feeling that he was not to be trusted.

Keeler left Lick Observatory in 1891 to become director of the Allegheny Observatory. This was not because of conflicts with Holden, but because he did not want to raise a family on the remote mountaintop. On the other hand, Burnham left in 1892 because he did not respect his boss. The most severe problems were between Holden and Barnard, which eventually led to Barnard's exit from Mt Hamilton.

Fig. 6.10. The 36 inch refractor of Lick Observatory, completed in 1888. The objective was made by Alvan Clark and Sons, the mounting by Warner and Swasey. James Lick, who donated $700 000 for the construction of the observatory, is buried in the pier of this telescope. When Lick was asked if it would not be easier to have his *cremated* remains placed in the pier of the telescope, he is reported to have said, 'No, sir! I intend to rot like a gentleman'. Photograph courtesy of the Mary Lea Shane Archives of the Lick Observatory.

When regular observing began with the 36 inch refractor, two nights a week were given to Burnham for double star observing, two were given to Keeler for spectroscopic work using the massive visual spectroscope he had designed, and two nights were used by Holden for photographic work and drawings of the Moon and planets. Saturday night was reserved for public viewing until about 11.00 PM, after which Burnham usually took over.[42] Barnard was not given any time on the telescope at all, having to content himself with the use of the 12 inch and the irregular gift of some of Burnham's 36 inch time, which the elder astronomer was happy to share on certain occasions of transitory phenomena such as eclipses of Jupiter's moons.

But this gift from a friend was not, however, the same as a regular right to some 36 inch time of his own. As a staff member Barnard felt he had this right, and was particularly infuriated with Holden because the director often stopped observing before the night was over, and Holden's results were not very good anyway. In fact, Holden was forever haunted by the reputation that he could not do scientific research. Back in 1877, shortly after Asaph Hall's discovery of the two moons of Mars, Holden thought he had discovered a third moon, the orbit of which 'not only violates Kepler's third law but also the simplest rules of geometry'.[43] Holden's magnum opus was a monograph on visual observations of the Orion Nebula, published in 1882. Many of Holden's photographs with the 36 inch at Lick were out of focus, and Barnard clearly could have done no worse with some of his own time. The issue came to a head in 1892 when the Lick Observatory visiting committee came to Mt Hamilton to investigate the bickering. Holden was forced to give Barnard some regular time on the 36 inch.

This produced an important result almost at once. In September of 1892 Barnard discovered the fifth moon of Jupiter, the first such discovery of a Jovian satellite since Galileo turned his little telescope skyward in 1609. Just as the discovery of Hyperion at Cambridge put Harvard College Observatory on the map and made the name of Bond famous, so did Barnard's discovery of Amalthea make him and Lick Observatory world famous. He was awarded the Gold Medal of the Royal Astronomical Society (RAS) in 1897, but by this time he was a staff member at the new Yerkes Observatory of the University of Chicago.

The short-lived high point of Holden's directorship was his acquisition of a 36 inch reflector from Edward Crossley, an upper class Englishman and one-time amateur astronomer. The telescope had been built by A. A. Common; with it Common had done some excellent astrophotography. His 1883

Fig. 6.11. The 36 inch Crossley reflector at Lick Observatory, as used by Keeler for his epochal photographs of nebulae, 1898–1900. Photograph courtesy of the Mary Lea Shane Archives of the Lick Observatory.

pictures of the Orion Nebula far surpassed the results of Henry Draper and contributed greatly to his being awarded the Gold Medal of the RAS in 1884. Common sold his telescope to Crossley and went on to build a 5 foot reflector. After 10 years, Crossley's interests had changed and he decided to sell the telescope. Not so much interested in the money, Crossley wanted his telescope to be situated where it could do useful astronomical research, something not possible in his native Yorkshire. Holden heard about the matter and thought that the telescope plus dome was a steal at £1150 (about $5750 at the time). However, Holden could not find a source of funds. Eventually Crossley decided to donate the telescope to the University of California, provided that they would pay for the shipment of the telescope and dome. It was estimated that this would cost $1000, an amount Holden easily obtained from wealthy San Franciscans. The telescope arrived in the summer of 1895, and Holden assigned to W. J. Hussey, primarily a double star observer, the job of putting 'the Crossley' (as it is still known) in order.[44]

Thus Lick Observatory was brought into accord with the original 1873 advice of Joseph Henry of the Smithsonian – to pursue positional astronomy with a first class refractor of the largest possible size, and to provide the option to pursue spectroscopic and photographic research by acquiring a large reflector. A reflector is particularly suited for this type of work because its focus does not depend on the wavelength of light. (The Lick refractor had a photographic correcting lens that could be mounted in front of the main objective. This brought the blue light to a better focus and, given that the photographic plates of that day were most sensitive in the blue, it led to adequate photographic results.) Also, a refractor and reflector of equal aperture are much different in cost – the reflector is much cheaper. Since the light gathering power (the ability to detect faint objects) is almost entirely a function of the area of the telescope objective, a properly set up shorter focal length reflector was just the thing for most astrophysical experiments. Not only does a shorter focal length telescope allow a smaller dome (costing much less), but the faster speed of the optics of the Crossley reflector (f/5.8 *v.* f/19 for the large refractor) gave brighter images, allowing the photography of faint objects with shorter exposures than required with the large refractor. While the Lick refractor had been used to take some excellent photographs of the Moon and planets, the Crossley reflector was better suited to photographing star clusters and nebulae, and did not require a photographic correcting lens.

I am getting ahead of the story a bit here. Hussey was unable and unwilling to expend the necessary engineering effort to make the Crossley work, particularly because of his desire to pursue his own research on the 36 inch refractor, and due to Holden's dictatorial manner. The result was that many old feelings of hatred resurfaced in the Lick staff, all but a couple of whom rallied to Hussey's defense. Holden was forced to resign and left Mt Hamilton in September of 1897, never to return.[45]

Simon Newcomb, Holden's mentor, writing in 1903, says, regarding Holden's administration, that

> ... [I]ts most singular feature was the constantly growing unpopularity of the director. I call it singular because, if we confine ourselves to the record, it would be difficult to assign any obvious reason for it. One fact is indisputable, and that is the wonderful success of the director in selecting young men who were to make the institution famous by their abilities and industry. If the highest problem of administration is to select the right men, the new director certainly mastered it. So far as liberty of research and publication went, the administration had the *appearance* [my italics] of being liberal in the extreme. Doubtless there was another side to the question. Nothing happens spontaneously, and the singular phenomenon of one who had done all this becoming a much hated man must have an adequate cause.[46]

Many of the successes of the 10 years of Holden's directorship were made in spite of him, rather than because of him. Burnham made many discoveries and measurements of double stars. Keeler measured emission lines in the spectra of nebulae and demonstrated that more definitive and more accurate work could be done with the 36 inch refractor at Mt Hamilton than the first generation of astrophysicists could do with their smaller telescopes under poorer seeing conditions in Europe. Barnard's discovery of the fifth moon of Jupiter certainly transpired in spite of Holden's lack of support. Schaeberle discovered a faint companion of Procyon in 1896, ending speculation of the existence of such a companion that arose in 1844 when Bessel proposed the idea to account for anomalies in Procyon's motion. This discovery could only have been made with a telescope like the Lick refractor. It testified to the excellence of the site and telescope, not to the greatness of the director.

Lick Observatory was the first permanent mountaintop observatory. It takes only a moment's thought to realize that it must be a constant psychological challenge to live and work at such an isolated site, especially when one considers the constraints on one's social and family life. If a director is supposed to provide a good example of technical and scientific knowhow and to marshal his troops to the best of their abilities to pursue scientific endeavors while soothing any ruffled feathers, then Holden was clearly not the right person for the director's job. His successor, Keeler, was.

In the 7 years that Keeler had been away from Lick he had matured as a scientist and gained valuable experience in the administration of observatory matters. He and Hale had founded the *Astrophysical Journal.* He had become world famous for his demonstration in 1895 of the true structure of the rings of Saturn. Using the Allegheny 13 inch refractor Keeler obtained a photographic spectrum of the planet and the rings which clearly showed that while the ball of the planet rotated as a single body, the rings must be composed of innumerable small particles. The orbital velocity of the rings decreased outward in accord with Kepler's third law; if the rings were a solid body the opposite would have been the case. This finding confirmed a theoretical result elaborated in 1859 by J. C. Maxwell, and while it was a simple and direct demonstration of the Doppler principle which could have been made by many people on other comparably sized telescopes, Keeler did it first. Science does not have to be complicated to be first rate.

When Keeler returned to Lick the staff consisted of three senior astronomers (W. W. Campbell, W. J. Hussey and R. H. Tucker) and three junior astronomers (C. D. Perrine, W. H. Wright and R. G. Aitken). (Schaeberle left Mt Hamilton in a huff because he had been acting director after Holden's departure, but was not chosen to succeed him permanently. Instead of finding a replacement for Schaeberle, Keeler used that salary money to create three fellowship positions for graduate students to gain practical experience at the observatory.) Campbell, assisted by Wright, was busy taking photographic spectra of all stars brighter than magnitude $5\frac{1}{2}$ in order to determine stellar radial velocities (motions along the line of sight). This project used half the time on the 36 inch refractor. Tucker was a meridian circle observer. Perrine started out as Lick secretary under Holden and became an astronomer in his own right. R. G. Aitken became one of the leading double star observers of the twentieth century and worked at Lick for over 40 years. These men were all busy with their own research and were more than happy when Keeler announced that he would personally take charge of the Crossley and drum it into shape.

For the reasons outlined earlier, Keeler knew that the Crossley could become quite useful for spectroscopic and photographic work. To begin with he had to design a new wedge-shaped base for the telescope which allowed its polar axis to be properly aligned on the north celestial pole. It was necessary to improve the drive system and the means of guiding on a star while taking a photograph. One of the features of the Crossley was that instruments were mounted at the prime focus while one guided at the Newtonian focus (at the side of the top of the tube). Not only did this place the observer precariously high in the dome, but it made the telescope difficult to balance; for every pound placed at the top of the tube, 10 to 20 times as much had to placed at the other end. (All large modern reflectors have a Cassegrain arrangement, in which there is a large hole in the center of the main mirror and a convex

Fig. 6.12. Photograph of the Orion Nebula by James Keeler. 40 min exposure with the Crossley reflector at Lick Observatory, 16 November 1898. From the *Lick Observatory Bulletin*, vol. VIII, 1908.

secondary mirror near the prime focus which directs the light through the hole to a focus behind the main mirror. This allows for easier access to the equipment and makes the telescope easier to balance.) Other problems with the Crossley included the fact that the mirror occasionally slipped in its cell, and the mounting was insufficiently rigid to withstand moderate winds without shaking. Originally, the prime focus spectrograph, when mounted on the telescope, protruded through the dome slit. In July of 1898 the mounting was cut down 2 feet to eliminate that problem. When Keeler had taken over the Crossley he had found it to be 'practically unmanageable by a single person'.[47] And, to this day, a successful observing run on the Crossley is indicative of one's level of practical ingenuity. The Crossley observing books are a kind of who's who of observational astronomy, for anyone who could obtain data successfully on this instrument would probably clean up like a bandit on another, better designed, telescope.

A description of Keeler's endeavors to make the Crossley useable, along with 70 of his pictures of nebulae and star clusters, were posthumously published in volume VIII of the *Publications of the Lick Observatory* in 1908. His prime conclusions resulting from his efforts were: (*a*) many thousands of unrecorded nebulae exist in the sky; Keeler estimated the number of nebulae accessible to the Crossley as 120 000; this number was later revised upward by Perrine to 500 000; in the 1930s N. U. Mayall estimated that $4\frac{1}{2}$–5 million nebulae could be photographed by the Crossley if there were no galactic obscuration; (*b*) many of the nebulae were spiral in nature. It was Keeler's results from the Crossley that led others like H. D. Curtis and H. Shapley to unravel the nature of the spiral nebulae, now known to be external galaxies. While long-focus refractors are still very useful for positional astronomy, for astrophysical research Keeler showed that reflectors were viable, and 'no American professional astronomer ever thought seriously of building a very large telescope as anything but a reflector, after Keeler's work with the Crossley'.[48] Unfortunately for Lick Observatory, Keeler's tenure as director was brief. He died in 1900, a month before his 43rd birthday.

By the turn of the century Lick Observatory had made a name for itself, and, like Harvard, sought to extend its researches to the southern hemisphere. Lick astronomers organized the D. O. Mills expedition and in 1903 established a station at Cerro San Cristobal near Valparaiso, Chile. The observatory there contained a $36\frac{1}{2}$ inch Cassegrain reflector, which was used for taking photographic spectra of southern stars. Lick operated this station until 1928. The Lick radial velocity program at Mt Hamilton and in Chile went on for a total of 35 years, resulting in over 25 000 spectra of 2770 stars.[49,50]

The Crossley reflector was redesigned in 1902–4 by Perrine, giving it the 'cement mixer' appearance it has to this day. Perrine was then its prime user and discovered the sixth and seventh moons of Jupiter with it photographically in 1904–5. (The ninth satellite of Jupiter was discovered with the Crossley by S. B. Nicholson in 1914.) Mostly, however, Perrine continued Keeler's work on photographing nebulae until he left Lick to become director of the Argentine National Observatory in 1909. H. D. Curtis was the principal Crossley user in the 1910s. We have already discussed his efforts to determine the nature of external galaxies and his participation in the Great Debate of 1920.

One of the epochal discoveries of the twentieth century was made by R. J. Trumpler, based on observations with the Crossley and the 36 inch refractor. Trumpler was studying open star clusters. He crudely assumed that they were the same physical size, in which case their angular diameters were indicative of their distances. (The smallest ones were the most distant. They would also have the faintest stars.) On the basis of the brightness of the stars in the clusters he showed that the photometrically derived distances differed from the geometrically derived distances more and more systematically as the clusters got fainter and smaller. This was the first hard evidence that there existed a general distribution of absorbing material in our galaxy, which dimmed the stars in an accumulating fashion as they became more distant. Other, not quite so hard evidence had been accumulating for some time, beginning with the star gauges of W. Herschel and W. Struve, and the reasoning by analogy, based on Crossley photographs by Keeler and Curtis, that if spiral nebulae show bands of obscuration and are galaxies in their own right, then our galaxy should also contain obscuring matter. After Trumpler's classic paper on interstellar absorption was published in 1930, no one ever doubted its existence again.

In contrast to Harvard, the equipment at Lick has not grown in number by leaps and bounds. In the 1940s Lick observatory added the Carnegie double astrograph, which is used to take wide field photographs simultaneously in blue and yellow light. These photographs are used to derive positional information on stars and solar system objects (planets, asteroids, comets) with respect to distant galaxies. Previously, parallaxes and proper motions of stars were derived with respect to faint stars within our galaxy, but because a star of faint apparent magnitude could be a nearby low-luminosity star, this could lead to errors. Using faint galaxies gives a much more fixed frame of reference.

One of the most important projects carried out with the double astrograph was the program of galaxy counts by Shane and Wirtanen, carried out from 1947 to 1964. Their photographic survey covered the whole sky down to declination $-23°$. They registered 1 250 000 images, of which 800 000 were galaxies. The photographic magnitude limit was 18.8. The Shane and Wirtanen counts are a prime source of information on the large scale structure of the universe, leading to the modern notions of superclusters of galaxies and bridges between the superclusters.[51]

In addition to a 24 inch photometric reflector, the Lick Observatory arsenal is rounded out by the 120 inch Shane reflector, which went into operation in 1959. It was the world's second largest telescope at the time of its completion. (Presently, it is at least twelfth.) It is used for optical and infrared astronomy with state-of-the-art instrumentation such as the 'Wampler scanner', an image tube spectrometer which provides real-time spectral energy information on the object under study, without the need for photographic plates and photographic reduction procedures.[52] Results from

Fig. 6.13. The 3 m Shane reflector at Lick Observatory, which went into operation in 1959. It was then the second largest telescope in the world. Photograph courtesy of the Lick Observatory.

the Shane 3 m reflector fill the pages of the *Astrophysical Journal* and deal with all aspects of modern astrophysics, from lunar laser ranging to emission lines in quasars.

As Lick Observatory approaches its 100th birthday in 1988, it is still a prime force in positional astronomy and astrophysics. Its role as an educational training ground for young astronomers also ranks among the highest. This began with the first Lick graduate student, A. O. Leuschner, in 1890, continuing through the 1960s with the Lick fellowships, organized by Keeler in 1898.[53] Following the inauguration of the Santa Cruz campus of the University of California (UC) in 1965, the headquarters of Lick Observatory were moved there from Mt Hamilton. UC Santa Cruz may very well be the most beautiful university campus in the world. There one may wander through the redwood-filled rolling hills, enjoy the clean, crisp, California air and behold the blue beauty of the Pacific and the outline of distant hills at the other end of Monterey Bay.

Of the many areas of research carried out at Lick, one we have not mentioned is solar astronomy. In addition to going on solar eclipse expeditions, for a number of years photographs of the Sun were taken at Lick with a photoheliograph, an instrument invented by George Ellery Hale in 1889 while still a Massachusetts Institute of Technology (MIT) undergraduate. It is used to obtain a picture of the Sun in a very narrow wavelength range such as that centered on a particular absorption line in the solar spectrum (a hydrogen or calcium line), which gives a somewhat hairy-looking, high-contrast image of the prominences over the whole face of the Sun. This program at Lick was discontinued after the establishment of the Mt Wilson Observatory (p. 165),[54] also the creation of G. E. Hale. But before we get to Hale's ultimate entrepreneurial accomplishments, we must first discuss the third member of the triumvirate of American astrophysics of the nineteenth century, Yerkes Observatory, which was Hale's first great success.

In 1890 Hale graduated from MIT, got married 2 days later, and spent part of his honeymoon at Lick Observatory. He then began his serious work in solar astronomy at the Kenwood Observatory, which was funded by Hale's father and situated in the backyard of the family house at 4545 Drexel Boulevard on the south side of Chicago. With his 12 inch refractor by Brashear and a spectroheliograph Hale photographed the Sun in the lines of calcium. These were the basic tools of early astrophysics: telescope plus spectrum-producing device plus photographic plate. Following in the footsteps of such men as Huggins and Lockyer, Hale was a pioneering amateur, but soon he was called to a university career as well. The University of Chicago was founded in 1892 by John D. Rockefeller. Its first president was the brilliant William Rainey Harper. Hale was appointed assistant professor of astronomy with the understanding that the Kenwood Observatory would become part of the University. But this transfer of the family observatory to the University of Chicago was contingent upon G. E. Hale's being given a permanent job, and also contingent upon the University raising $250 000 for a larger observatory. The dreams of a major observatory were bolstered that very year (1892) when it became known that the Clarks had blanks large enough to make the objective for a 40-inch refractor. At about the time Lick Observatory was being finished, a group at the University of Southern California made plans for the establishment of a larger telescope near

Pasadena, but that did not come to pass and the Clarks were eager to find someone else to pay for the lenses.

Though Hale taught no formal classes at the university, he was kept in charge of the Kenwood Observatory and was to supervise solar astrophysics work. His main task was to plan for the eventual new observatory. A Chicago millionaire, Charles Tyson Yerkes, who had made his money building streetcar systems, agreed that an observatory could well perpetuate his name, similar to the plans of James Lick for perpetuating the Lick name. In the fall of 1892 Yerkes agreed to provide the funding for the largest telescope in the world, for a cost estimated to be half a million dollars. In the early going money was no object, as long as they could 'lick the Lick'.

One of the prime considerations was where to place the observatory. Hale was primarily interested in solar work, and his brief experience at Mt Hamilton had shown him that the daytime conditions are not necessarily good at a site which had good nighttime seeing. He tells us:

> If there had been absolute freedom of choice, a site combining the excellent conditions for night work at Mt Hamilton with the good day seeing existing elsewhere would have been sought far and wide, without regard to geographical boundaries.[55]

They considered a number of places in the Chicago suburbs, a few sites in rural northwest Illinois, Lake Geneva, Wisconsin and Pasadena, California, the originally intended site for the University of Southern California (USC) telescope. Advice on the site selection criteria was solicited from Barnard, Burnham, Keeler, Pickering, Newcomb and others. Both the Board of Trustees of the University of Chicago and Yerkes himself wanted the observatory to be located within 100 miles of the city of Chicago, thus eliminating Mt Wilson near Pasadena once again. The site chosen was Williams Bay, Wisconsin, on the northern shores of Lake Geneva.

At the Columbian Exposition in Chicago in 1893 the mounting and tube of the Yerkes telescope were put on display (and at one point they were very nearly damaged in a fire). An international astronomical congress took place that August in connection with the Columbian Exposition.[56] In Chicago enthusiasm was high for the new telescope, its observatory, and astronomy in general.

The lenses for the Yerkes telescope were finished in the summer of 1895 by the Clarks and tested in Cambridge by Hale and Keeler. Their view of the Orion Nebula made them rightly believe that the Yerkes telescope would indeed be the most magnificent refractor in the world.

The construction of the observatory was proceeding, albeit too slowly for

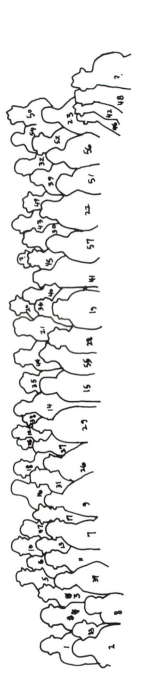

Hale's liking. By the end of 1896 the Kenwood dome and 12 inch telescope had been moved to Williams Bay, and Hale and his family were living there too.

In May of 1897 the lenses were finally shipped to Lake Geneva and mounted in the more than 60 foot long tube. Further testing, alignment, and engineering was begun at once, but at the end of the month a major calamity befell the observatory. The cables which held the massive elevator-type floor in the dome of the great refractor had snapped, and the inside of the dome looked 'as if a cyclone had slipped through the slit and gone on a rampage'.[57]

The telescope was finally dedicated on 21 October 1897. Over 700 people were present for the dedication.[58] The keynote address was given by Keeler, who was by now recognized to be the most promising young astrophysicist in America besides Hale. The title of his talk was 'The importance of astrophysical research and the relation of astrophysics to other physical sciences'.[59] He stressed the importance of the spectroscope in astrophysics for determining stellar radial velocities, the discovery of spectroscopic binaries, the measurement of magnetic fields in celestial bodies (i.e., stars, including the Sun), and also the importance of photography for determining the

Fig. 6.14. An assembly of some of the guests at the 1897 dedication of the Yerkes Observatory. Identifications made about 1 January 1923 by F. Ellerman, E. E. Barnard, Storrs B. Barrett, JAP.

1. E. E. Barnard, 2. C. H. Rockwell, 3. Geo. F. Hull, 4. Colton (?), 5. Kurt Laves, 6. Frank W. Very, 7. E. B. Frost, 8. Henry M. Paul, 9. Ernest Fox Nichols, 10. F. R. Moulton (?), 11. Ephraim Miller, 12. Father John Hedrick, 13. John M. VanVleck, 14. Milton Updegraff, 15. Wm R. Brooks, 16. F. L. O. Wadsworth, 17. H. C. Lord, 18. F. H. Seares, 19. Geo. W. Hough, 20. W. H. Collins, 21. Caroline E. Furness, 22. Mrs Pickering, 23. Mrs Hale, 24. Miss Cunningham, 25. Alva A. Lyon, 26. Carl A. R. Lundin, 27. J. A. Parkhurst, 28. John A. Brashear, 29. G. W. Ritchey, 30. C. H. McLeod, 31. Father J. G. Hagen, 32. Charles Lane Poor, 33. J. K. Rees, 34. Miss Mary W. Whitney, 35. F. P. Leavenworth, 36. Henry S. Pritchett, 37. Jas. E. Keeler, 38. A. G. Stillhamer, 39. Hugh L. Callendar, 40. Geo. W. Myers, 41. Chas. L. Doolittle, 42. E. C. Pickering, 43. A. W. Quimby, 44. Asaph Hall Jr, 45. Albert S. Flint, 46. M. B. Snyder, 47. W. W. Payne, 48. Carl Runge, 49. Winslow Upton, 50. Geo. Kathan, 51. G. D. Swezey, 52. Geo. E. Hale, 53. N. E. Bennett, 54. Geo. C. Mors, 55. F. Ellerman, 56. W. J. Humphreys, 57. Henry Crew

In Guestbook, but not in photograph: A. C. Behr, E. Colbert, W. Harkness, L. H. Ingham, T. S. Smith, E. E. Lockridge, Wylczynski, M. McNeill, E. A. Engler, Carhart, Moran, Thwing, Comstock, Todd, Rotch.

The First Conference of American Astronomers: The American Astronomical Society was organized at the 1897 dedication, and the first regular meeting of the Society was held at the Yerkes Observatory on 6, 7 and 8 September 1899.

Yerkes Observatory photograph, University of Chicago.

morphology of nebulae, comets, and the atmosphere of the Sun. The advantage of combining these two methods of the new astronomy to obtain photographic spectra was self-evident.

In his talk Keeler said:

> Those who are interested in the results of science, but who care little for methods, and know nothing of elegant forms of analysis, are naturally more attracted by the view of the heavenly bodies which astrophysics presents, than by the view which is obtained from the standpoint of the older astronomy.[60]

By this he meant that it requires quite a great deal of mathematical expertise to compute orbits or determine the perturbations of various planets on each other, but it can easily be understood from photographs or spectra something of the structure and composition of celestial bodies which cannot be determined through the methods of positional astronomy. Keeler's own demonstration of the many-particle nature of the rings of Saturn is a case in point.

Not only was George Ellery Hale a great observatory builder, but he was interested in the proper publication of astronomical results (as we have already seen with the *Astrophysical Journal*), and he was also interested in the exchange of ideas at scientific meetings. The coming together of so many notable astronomers at the Yerkes dedication led directly to the formation of the American Astronomical Society.[61]

The primary research envisioned for Yerkes Observatory had already been outlined by Hale in 1892:

> ... solar observations, comprising visual and photographic studies of the structure of the photosphere, spots and faculae, photography of the faculae, chromosphere and prominences with the spectroheliograph, spectroscopic observations, both visual and photographic, of all classes of solar phenomena, and bolometric and photometric investigations of various kinds; micrometric observations of double stars, nebulae, planets, satellites, comets, etc.; photography of the Milky Way, stars, nebulae, etc.; researches in stellar spectroscopy; meridian observations [for time determinations primarily]; laboratory work of various kinds, principally with the spectroscope, bolometer and refractometer.[62]

In addition to the 40 inch refractor and the 12 inch refractor, the other equipment at the new observatory included a transit instrument, a 10 inch astrograph funded by Catherine Wolf Bruce, the same woman who had funded Harvard's 24 inch photographic telescope used in Peru, and a 24 inch photographic reflector built by George W. Ritchey, an assistant of Hale's

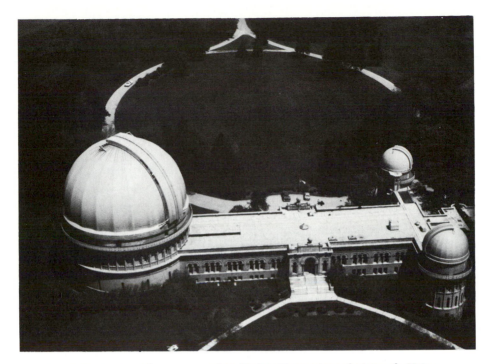

Fig. 6.15. Yerkes Observatory today. Yerkes Observatory photograph, University of Chicago.

at Kenwood. In addition to Hale and Ritchey the scientific and technical staff consisted of E. E. Barnard and S. W. Burnham, both formerly of Lick Observatory, Ferdinand Ellerman (another of Hale's Kenwood assistants), and F. L. O. Wadsworth, who had been an assistant of Michelson's at the University of Chicago. Hale had tried to hire Keeler to be in charge of nighttime spectroscopic work, but Keeler was busy trying to improve the state of astronomy facilities in the Pittsburgh area and shortly after his big splash as the main speaker at the Yerkes dedication was offered the job of Lick director. One of the most significant additions to the Yerkes staff came in 1898 when Hale hired Walter S. Adams, who would become Hale's right-hand man at Yerkes, then subsequently at Mt Wilson.

Just how good was the Yerkes 40 inch refractor, then the largest telescope in the world? Compared to the Lick 36 inch, the Yerkes telescope had 23% greater light-gathering power. It could theoretically see fainter objects and separate more closely spaced double stars, though in the latter case the Earth's atmosphere would in general wipe out any distinctions of resolving power. Given that the lenses and mountings of the two telescopes were made by the same respective manufacturers, the Yerkes telescope was expected to surpass the 36 inch. Barnard and Burnham both made that very assessment,[63] but this must be interpreted as an accomplishment on the part of the Clarks to produce a still larger objective of the same excellent workmanship as the 36 inch, rather than indicating any flaws with the Lick telescope. As for the site, Hale wrote: 'Professor Barnard has found that the best nights [at Williams Bay] are fully as good as the best nights at the Lick Observatory, though the average night seeing is not as good as it is at Mt Hamilton'.[64] The daytime seeing at Yerkes was far superior to that at Lick, which was of prime importance for Hale's solar research. In contrasting the two sites, one must also consider the seasons. While a clear night at Yerkes may rival a clear night at Mt Hamilton, at the latter there are many more such clear nights in a year, especially from April to October when Lick can almost guarantee good weather to the observers. The wintertime at any observatory is not as good, but at least at Lick it is not as cold as southern Wisconsin, where the temperature may fall to $-25\,°F$ ($-32\,°C$).

In the eye of the public the prime distinguishing feature of Yerkes Observatory is that the 40 inch refractor is still the largest refractor in the world. It is likely to remain so, for it is so much cheaper to provide the equivalent light-gathering power with a reflector, and astrometric reflectors, such as the Naval Observatory's 61 inch situated at Flagstaff, Arizona, have shown that positional astronomy need not necessarily be carried out with a long-focus refractor. Yet, for some types of work, the Yerkes telescope remains unsurpassed because of its long focal length (over 19 m). Its lenses have never been removed from their cell, thus allowing a direct comparison of star positions obtained over a span of 90 years without having to worry about some sources of systematic errors. Only a dozen or so telescopes in the world can be used effectively for the determination of stellar parallaxes, and the Yerkes telescope still reigns as the king. In addition to the long-term projects for the measurement of parallaxes, proper motions, and double star positions, the Yerkes telescope is also used for studies of the dynamics of star clusters. A 24 inch telescope at Yerkes (not the original one) is still in use, primarily for photoelectric photometry. More recently (mid-1960s) it has been joined by a 41 inch reflector, which is used for spectroscopic studies.

Many distinguished astronomers have been associated with the Yerkes Observatory, among them: W. W. Morgan, who has been at Yerkes for 60 years and has had a prime responsibility for our modern stellar classification scheme, the Morgan–Keenan system, and also for the *UBV* system of broad band optical photometry;[65] Otto Struve, director of Yerkes from 1932 to 1947, known for his studies of stellar rotation and stellar evolution, his great prolificacy (six books and nearly 700 articles), as well as his famous family name;[66] Gerard P. Kuiper, director of Yerkes between 1947 and 1960, who

Fig. 6.16. The night sky brightness of the Midwestern region of the central United States, 1979. Yerkes Observatory is too close to the Chicago metropolitan area to enjoy dark skies. From USAF DMSP (Defense Meteorological Satellite Program) film transparencies archived for NOAA/NESDIS at the University of Colorado, CIRES/National Snow and Ice Data Center.

discovered the fifth moon of Uranus in 1948 and the second moon of Neptune in 1949, and who was a prime mover in establishing other astronomical institutions, such as the Lunar and Planetary Laboratory of the University of Arizona and the Mauna Kea Observatory in Hawaii (p. 225ff.); and S. Chandrasekhar, who has been associated with Yerkes and the University of Chicago for over 50 years. He is the foremost astrophysical theorist in the world and the co-recipient of the 1983 Nobel Prize in physics.

The desire for bigger telescopes and better seeing conditions led Hale to California. He brought Ritchey and W. S. Adams with him. Back in 1896 Hale's father had donated a 60 inch mirror blank to the University of Chicago. Ritchey worked on the mirror at Yerkes, but the telescope came to be located at Mt Wilson.

Yerkes astronomers themselves opened an auxiliary station. From 1932 to 1962 the Yerkes director was also director of the McDonald Observatory on Mt Locke in Texas. It has operated an 82 inch reflector since 1939, and since 1969 has also had a 107 inch reflector. The astronomy program at the University of Texas, which now takes care of McDonald without the Yerkes connection, is one of the major programs in the country.

The seeing at Yerkes is still good enough for much astrometric and spectroscopic work, but the nighttime sky is much brighter than it was 90 years ago because of the growth of the Chicago metropolitan area (see Fig. 6.16). But, not to despair, one of the strong programs at Yerkes is infrared astronomy, carried out on the Kuiper Airborne Observatory (based at Moffett Field, California – see p. 254) under the direction of the present director of Yerkes, D. A. Harper. The Yerkes infrared group also observes at Mauna Kea in Hawaii and has observed at the South Pole.

Harvard, Lick and Yerkes clearly remain astronomical capitals of the world. A doctorate earned in connection with any of them has been proven to be a pretty accurate indicator of a high level of scientific ability and a good predictor of success as a research scientist. To describe all the research carried out at these three institutions and list all the important individuals associated with their famous telescopes would fill much more than three books. All three of these observatories have expanded far beyond their original environs. Harvard and Lick operated southern hemisphere operations. Harvard became a center for satellite-borne X-ray astronomy and also it plays a major role in the operation of the Multiple Mirror Telescope. Lick and the University of California sought a darker site than Mt Hamilton and hoped to establish another station at Junípero Serra Peak in California. More recently, with Caltech, they instead became committed to establishing a 10 m telescope at Mauna Kea in Hawaii (see p. 267–8). The University of

Chicago operated McDonald Observatory for many years and is now part of a consortium of universities involved in building a 3.5 m reflector in New Mexico.

The involvement of Harvard, Lick, and Yerkes in astronomical research continues to be pathbreaking and exciting. We can expect their future to be as brilliant as their past and present.

7

Mt Wilson and Palomar

Like buried treasures, the outposts of the universe have beckoned to the adventurous from immemorial times. Princes and potentates, political or industrial, equally with men of science, have felt the lure of the uncharted seas of space, and through their provision of instrumental means the sphere of exploration has rapidly widened ... Each expedition into remoter space has made new discoveries and brought back permanent additions to our knowledge of the heavens. The latest explorers have worked beyond the boundaries of the Milky Way in the realm of 'island universes', the first of which lies a million light-years from the earth while the farthest is immeasurably remote. As yet we can barely discern a few of the countless suns in the nearest of these spiral systems and begin to trace their resemblance with the stars in the coils of the Milky Way. While much progress has been made, the greatest possibilities still lie in the future.
George Ellery Hale (1928)

To prove that the Mt Wilson and Palomar Observatories are important is much like proving that $2 + 2 = 4$. It is something everybody knows already. The 100 inch telescope at Mt Wilson, which saw its first light on 2 November 1917, was the most significant telescope in the world for 30 years. It opened up the doors of the universe almost singlehandedly. Extragalactic astronomy, born at Lick and Lowell Observatories, reached maturity at Mt Wilson and Palomar. The 100 inch was only surpassed when the 200 inch at Palomar came on line in 1948, which in turn became the most significant telescope in the world at least for its first 30 years and which is undeniably the most famous telescope ever built. Pasadena, California, the headquarters of these two institutions, is still regarded by many to be the site of the Number One Club, and with the construction of the Caltech/University of California 10 m Keck telescope for optical and infrared astronomy and the Caltech 10.4 m submillimeter wave telescope, both to be placed at the summit of Mauna Kea in Hawaii (see p. 267ff.), Pasadena will certainly remain a major astronomical center of the world.

The two institutions covered in this chapter have had a variety of official names. Before there was any official permanent observatory at Mt Wilson, George Ellery Hale's next brainchild was known as the Mt Wilson Station of the Yerkes Observatory. In 1904, with the official founding of the observatory, it became the Solar Observatory of the Carnegie Institution of Washington, which meant the same as Mt Wilson Solar Observatory. In 1920 it became simply the Mt Wilson Observatory. In 1948, with the completion of the 200 inch at Palomar Mountain (there is no such place as Mt Palomar), the observatories were known as the Mt Wilson and Palomar Observatories. In December of 1969 they were renamed the Hale Observatories. In 1980 they became two separate institutions, the Mt Wilson and Las Campanas Observatories, and Palomar Observatory. The former has been funded by the Carnegie Institution, while the latter is funded by the California Institute of Technology. Both of these great observatories were the progeny of George Ellery Hale, and while the younger 'brother' was destined to become more powerful and the sibling rivalry between the two led to their going separate ways, no one can deny their common heritage. The Mt Wilson and Palomar Observatories are permanently linked in our minds.

We have already mentioned how the University of Southern California had wished to establish a 40 inch refractor on Wilson's Peak (as it was then called), a 5703 foot mountain at the south end of the Sierra Madre range. This was in response to the establishment of Lick Observatory. In 1888 the President of the University of Southern California, the Rev. Dr M. M. Bovard, had paid a visit to E. C. Pickering at Harvard to discuss the matter. Pickering suggested that a 24 inch photographic telescope would be a better idea, but Bovard insisted that southern California have a telescope larger than the Lick telescope. These two men agreed over the next year to pursue a joint venture at Wilson's Peak. Observations were carried out there and photographs were taken with a Clark 13 inch refractor, but there were many problems. Getting equipment to the mountain was difficult. The 9 mile trail out of Pasadena was only 2 feet wide in many places. Equipment had to be brought up by mule train. Living conditions at the summit were primitive. There was heavy snow in the winter. The isolation and intensity of the work was bound to lead to bad tempers at the mountain. To make matters worse for the establishment of an observatory at that time, there arose a rivalry between Pasadena and Los Angeles and suspicions of what Harvard's role would be in the running of the observatory. There were controversies over who owned the mountain—the Southern Pacific Railroad, the Federal Government, or the State of California. There were legal entanglements, disputes in Congress, distrust on the part of southern Californians of everyone in northern California, Washington and Cambridge. And the land boom in the Los Angeles area which was counted on to finance the USC observatory went bust. Harvard went on to establish its station in Peru, and the disks for the USC telescope ended up as the objective of the Yerkes refractor.[1]

Mt Wilson, as it has been referred to since the turn of the century, was nevertheless destined to have a permanent observatory.[2,3] In 1902 the Carnegie Institution of Washington was founded, whose purpose was to fund worthy educational and peaceful causes. Their original endowment was $10 million. During its first year of existence S. P. Langley of the Smithsonian wrote to the Carnegie Institution, urging them to establish a high altitude solar observatory. A committee was set up, consisting of Langley, E. C. Pickering, Lewis Boss, Newcomb and Hale. This led to the establishment of a second committee, which was to draw up a detailed plan for solar astronomy, as well as a plan for a southern hemisphere observatory. The second committee consisted of Lewis Boss, W. W. Campbell (the Lick director) and Hale. The report of the second committee included the results of site testing by W. J. Hussey of Lick, who had investigated the atmospheric conditions in southern California and Arizona, and who recommended Mt Wilson first and foremost.[4]

Hale's first trip to the mountain was in June of 1903. His previous experience at other mountains had not prepared him for the superiority of the seeing conditions at Mt Wilson. He found the solar image to be almost perfect. This first trip was of short duration. He decided to make more extensive tests. He wrote:

> As circumstances required that my family should spend the winter of 1903–4 in southern California, I decided to take this opportunity to make a more complete test of atmospheric conditions on Mount Wilson. Before arrangements had been made for living upon the mountain, I made frequent trips from Pasadena to Mount Wilson during the months of December, January, and February, observing the Sun on each occasion with a telescope of $3\frac{1}{4}$ inches aperture, and noting the prevailing weather conditions. The extraordinary absence of wind, which had seemed so characteristic a feature of the mountain during Professor Hussey's visit, could not be said to continue throughout the winter months. High gales sometimes occur at this season, and the average wind velocity is greater than during the summer. Nevertheless, the wind during the day was usually very light, and on many occasions the quiet days of the previous June seemed to be almost exactly duplicated, except that the temperature was lower. For weeks together not a cloud would be seen in the sky, and the summer serenity was in some measure retained until well into January. Later it was broken by storms, but these practically ended with April.

> As the solar definition proved to be surprisingly good for this season of
> the year, I was soon convinced that Mount Wilson offered exceptional
> opportunities for both solar and stellar work and that a systematic test of
> conditions should be inaugurated at the earliest possible moment.[5]

Systematic records of the cloudiness, humidity, temperature, atmospheric
pressure, atmospheric transparency, and seeing stability were begun in April
of 1904. Hale wrote:

> ... it appears that Mount Wilson meets in a very remarkable degree the
> requirements of a site for a solar observatory. Indeed, I know of no other
> site that compares at all favorably with it. If a large solar observatory
> were established there, it might be expected to yield many important
> results, not to be obtained under less favorable conditions.[6]

The first major instrument installed at Mt Wilson was the Snow telescope,
which was officially on loan from Yerkes. When Hale was designing Yerkes
in 1894 he made a provision for a heliostat room, which would employ a flat
mirror, or coelostat, to reflect the Sun's light into a fixed solar telescope. One
sets up the coelostat in equatorial celestial coordinates by adjusting it in one
direction to match the declination of the Sun, and by driving it in the right
ascension direction by means of a motor to compensate for the Earth's rate of
rotation. After the initial experiments with two smaller instruments, Hale
made plans in 1900 to build a solar telescope containing a 30 inch diameter
coelostat and a plane mirror of 24 inches aperture which fed the sunlight
either to a concave mirror of 61 feet focal length or to a second concave
mirror of 165 feet focal length. The Yerkes heliostat room was only 104 feet
long, so a long horizontal wooden building was built to the south of the main
building. Small grants totalling $1000 saw to the construction of the
building and telescope, but it unfortunately burned down in December of
1902. A gift of $10 000 from Miss Helen Snow provided enough funds to
rebuild the instrument and place it in a more suitable building. The coelostat
was to be mounted 15 feet above the ground, as opposed to the previous
arrangement in which it was only inches from the ground. Already Hale was
finding that the higher the coelostat mirror was, the more stable the solar
image. The Snow telescope was completed in the autumn of 1903.

George Ellery Hale had the particular characteristic that as soon as he
reached the edge of a new scientific plateau, he would think about the climb
to the next one. No sooner was the Snow telescope operational at Yerkes
than he was off to California as part of the Carnegie Institution's committee
on the future of solar astronomy and planning on getting the Snow telescope
to Mt Wilson. The Carnegie Institution came up with $10 000 for
support of Hale's California plans, but this did not even match the amount

Fig. 7.1. The building for the Snow solar telescope at Mt Wilson. From *contributions from the Solar Observatory, Mt Wilson, California*, No. 2, 1905.

Hale had spent out of his own pocket. Later that year, on 20 December 1904, the Executive Committee of the Carnegie Institution voted in favor of a gift of $150 000 for each of 2 years. A permanent observatory at Mt Wilson was thus founded and funded.

Hale's first plan was to build a better solar telescope than the Snow telescope, even though he had just completed its move from Wisconsin to Mt Wilson. While carrying out his early tests in 1904, Hale had tested the seeing by making solar observations with a $3\frac{1}{4}$ inch telescope supported in a pine tree at heights ranging from 20 to 80 feet. He found that 'The results of these observations clearly indicate that a telescope employed in solar work should be mounted as high above the ground as circumstances warrant'.[7] Accordingly, a 60 foot tower telescope was planned. This had a 43 cm diameter coelostat, a 56 by 32 cm elliptical flat mirror, and a 30.5 cm objective doublet with a focal length of 18.3 m. While the coelostat was placed high above the ground and the associated air currents, the spectrograph for this new telescope was placed in a constant temperature room 30 feet underground, which provided for its thermal and vibrational stability. The 60 foot tower telescope, which became operational in 1907,

soon proved superior to the Snow telescope because of its vertical arrangement. But it did not have quite the same spectral resolution, because of its
shorter focal length. As a result a second, larger tower telescope was
planned, the 150 foot, which went into operation in 1912. It was a scaled up
and improved version of the 60 foot. The 60 foot produced a 17 cm solar
image, the 150 foot a 43 cm image. The tower of the 150 foot had double
walls, in essence a building within a building, to cut down any effects of the
wind.[8]

It was with the Snow telescope at Mt Wilson that Hale and his colleagues
first proved that sunspots were cooler regions of the Sun's photosphere. They
found that certain absorption lines in the solar spectrum were stronger when
the slit was directed at a large sunspot. These lines were observed in lower
temperature sources such as electric arcs measured in the laboratory, which
were known to be cooler than the general photospheric temperature.

It was with the 60 foot tower telescope that Hale made his greatest
discovery, the discovery that put Mt Wilson on the map, that of the presence
of strong magnetic fields in sunspots.[9] Previously, Zeeman had found from
laboratory measurements that a magnetic field along the line of sight splits a
spectral line into two circularly polarized lines. Hale wanted to investigate
this possibility in the Sun's spectrum, and in 1908 he observed that some
spectral lines known to be normally single in the laboratory were double if
one observed them in sunspots. In the lab he observed the spectra of sparks
under the influence of magnetic fields of 15 000 Gauss, and he estimated that
iron and calcium lines in the Sun's spectrum were made double by fields as
strong as 2900 Gauss. (The general field on the surface of the Sun away from
sunspots is only 1–2 Gauss.) As magnetic fields are caused by organized
motions of charged particles, like an electric current, Hale wanted to
determine the sign of the charge of the particles in the solar photosphere. He
determined that the strong magnetic fields were due to the flow of electrons
(which are negatively charged).

The most significant role of solar astronomy carried on at Mt Wilson has
been the assembling of a very large and consistent data base for solar
astronomers to work with. Since the founding of the observatory the Sun has
gone through more than 7 of its 11 year cycles of sunspots. Very long runs of
white light and calcium light images of the solar disk have been obtained. It
was with Mt Wilson data that Hale discovered the law of magnetic polarity of
sunspots. While the number of spots varies with an 11 year period, the
magnetic polarity (north or south) varies with a 22 year period. This means
that while every 11 years there will be many spots in, say, the northern
hemisphere of the Sun, it is only every other sunspot cycle that the spots will
have the same polarity.

Fig. 7.2. 'Truck' for hauling heavy instruments up the old Mt Wilson trail. From *Contributions from the Solar Observatory, Mt Wilson, California*, No. 2, 1905.

Hale's plan for the study of stellar evolution not only called for the observation of the Sun during the day, but also the observation of the stars at night. This of course required conventional telescopes, and those able to achieve as high a spectral dispersion as possible. This could only be done if the telescopes had as much light-gathering power as possible, hence those with large diameter objectives. Back in 1894, Hale's father agreed to buy the University of Chicago a mirror blank 60 inches in diameter, under the provision that the university would make a telescope out of it. The university did not uphold its part of the bargain, as there were no available funds. Hale's father did buy the mirror blank and hoped to provide the mounting as well. By 1898 Ritchey had finished the grinding of the spherical surface of the mirror at Yerkes,[10] but this was the year the elder Hale died, and the 60 inch project was put on hold.

It was not certain whose side the Fates were on regarding the construction of the 60 inch at Mt Wilson. Though the funding of the observatory portended a good omen, the fire which accompanied the 1906 San Francisco earthquake almost led to the destruction of the Union Iron Works; the mounting of the telescope had just been cast there. The winter of 1907 was extremely severe, and the mountain was sealed off for weeks at a time. The road had to be widened. Equipment was still being brought up by mule train or by means of special trucks (Fig. 7.2). By the end of November 1907, over 150 tons of material for building and dome had made it to the summit, and

Fig. 7.3. The 60 inch reflector at Mt Wilson in the early 1900s. Photograph courtesy of the Mt Wilson and Las Campanas Observatories.

by the following June the finished mirror was also at the top. The first photograph was taken with the 60 inch on 20 December 1908, 4 years to the day after the observatory's founding.[11]

After supervising the construction of the Yerkes 40 inch, the largest refractor in the world, Hale had now supervised the construction of the largest reflector, which not only had more than twice the light-gathering power of the Yerkes telescope, but was also situated at a decidedly better site. The glass for the 60 inch mirror had been cast at the Saint-Gobain glassworks in France. It was an 8 inch thick slab of greyish emerald-green plate glass weighing nearly a ton and figured by Ritchey to an f/5 curvature (a focal length of 25 feet). The telescope had a number of configurations. Photographs could be taken at prime focus, at Newtonian focus, or folded Cassegrain (f/20). A spectrograph could be mounted at the folded Cassegrain focus (f/16), or the light could be sent down the polar axis (the coudé configuration at f/30) to a massive spectrograph in the coudé room.

No sooner was the 60 inch reflector in operation than it began to produce epoch-making results. Such is to be hoped for from the newest, most advanced telescope, specially designed for the exploitation of an excellent site by a group of highly motivated scientists. Not everyone who tried to build and use reflectors had success, but with the success of the Crossley at Lick, the Yerkes 24 inch, and the Mt Wilson 60 inch, 'by 1909 there could be little doubt that in astrophysics the reflector had replaced the refractor as the primary observatory instrument'.[12]

Over the course of its lifetime, the 60 inch has fulfilled Hale's hopes that it would further astronomical knowledge. With it Ritchey and Shapley photographed novae in spiral nebulae. This was important for the subsequent proof that these nebulae were separate galaxies.[13] Shapley used the 60 inch a great deal to investigate variable stars in globular clusters, which, as we have already mentioned, contributed greatly to the modern view of the structure of our own galaxy.[14] Using the Cassegrain spectrograph of the 60 inch, W. S. Adams obtained spectra of faint stars, confirming the existence of white dwarf stars (very low luminosity, compact stars which are dying remnants of previously larger and usually more massive stars).[15] In 1914 Adams and Kohlschütter showed, from spectra taken with the 60 inch, that there is a correlation between certain spectral features and the absolute magnitudes of the stars, which allowed one to determine the distances to remote stars without their being associated with a star cluster, being Cepheids, or being close enough for a direct parallax determination.[16]

In addition to optical photography and spectroscopy the 60 inch has been used in some of the pioneering infrared work, notably by Eric Becklin and Gerry Neugebauer. In the late 1960s they mapped out the Galactic Center[17] and also the Orion Nebula.[18] The Becklin–Neugebauer object (or BN object, named after them), which they discovered in 1966 in the Orion Nebula, is a superluminous protostellar source 10 000 times brighter than the Sun. It will soon become a 10 solar mass star.

In recent years the 60 inch has been used for near nightly monitoring of the calcium emission in the spectra of cool stars (F, G and K Main Sequence stars) to investigate chromospheric activity and search for evidence of differential rotation.[19] This is exactly the kind of project that Hale would have liked to see get carried out, since it is an attempt to study other stars as Hale had studied the Sun. For this project the 'emission cores' of the Fraunhofer H and K lines are measured, from which one can derive information on the chromospheric activity of the stars. This project, headed up by Arthur Vaughan and Douglas Duncan and carried out by them and the rest of the K Flux Klan, has been allocated up to 360 nights of telescope time per year. Not only does it fulfill the original purpose of the observatory, to investigate stellar structure and evolution, but it can be carried out even though the sky at Mt Wilson is bright due to the lights of Los Angeles. While many astronomical projects can not be carried out in the face of light pollution or during the half of the month when the Moon makes even more light, the team at the 60 inch has worked happily on.

At this point we must mention something of the living conditions at the mountain. When Hale first arrived there, the only building was a cedar log cabin built in 1893 which had lain unused for the intervening years. It was known as the casino. At night one could see stars through the cracks in the roof.[20] During the summer of 1904 Hale and Adams set out one day to explore the ridge of the mountain extending southwest from the site of the Snow telescope. A quarter mile down from the ridge they found a level spot. 'On three sides the slopes fell abruptly into nearly sheer precipices, and the view of the valley, the canyons, and the distant mountains was magnificent and quite unobstructed... "This is where we must have the Monastery", [Hale said].'[21] Why call it a monastery?

> In the first place, it is by no means desirable to confine families, and especially children deserving every educational advantage, within the narrow limits of an isolated observatory colony. Furthermore, the work of an instrument shop, and much routine computing as well, can be done at much less expense and to better advantage in a town... The isolation of most mountain sites, however, might seem to demand that the staff of such an observatory should be composed only of celibates, or that its members must be content to experience long periods of separation from their families.[22]

So wrote Hale. After the goings on at Lick and Yerkes, there was a plain need to try this separation of work and family life. While still no picnic for the 'astronomy wives' to see their husbands on weekends only or between observing runs, this practice of keeping observatory peaks as places for work but not families has been practiced to this day, and it allows the astronomers to concentrate on their work without the normal distractions of home. The Monastery at Mt Wilson, completed in late 1904, contained seven bedrooms, four offices, a dining room, library, and kitchen. It was the principal residence at the mountain for many years.

The most powerful telescope at Mt Wilson, an engineering marvel of its time, the largest telescope in the world from 1917 to 1948, a scaled-up version with improvements (naturally) of the 60 inch, was the 100 inch Hooker reflector.[23] Even before Hale had seen the light at the end of the tunnel building the 60 inch, the photomaniacal entrepreneur was dreaming of a larger telescope. In 1906 he convinced a wealthy Los Angeles businessman, J. D. Hooker, to donate $45 000 for the mirror blank, which, like the 60 inch, was cast of green plate glass at the Saint-Gobain glassworks in France. The disk that was eventually used for the main mirror was the first one cast (in 1906), which was delivered to Pasadena in late 1908. The Mt Wilson staff was greatly disappointed with this blank for it contained a large number of embedded bubbles. They ordered the glassworks to perform some experiments and cast a better blank, but after subsequent attempts which were no more successful, Hale ordered Ritchey to begin work on grinding the first blank. The surface was ground to a spherical shape by the end of 1914 without having encountered any of the embedded bubbles. The mirror was then parabolized (to eliminate the effects of spherical aberration) by W. L. Kinney. First light was achieved on 2 November 1917.[24] Adams tells us:

> Soon after dark the telescope was swung over to the eastward and set on the planet Jupiter, and we had our first look through the great instrument. The sight appalled us, for instead of a single image we had six or seven partially overlapping images irregularly spaced and filling much of the eyepiece. It appeared as if the surface of the mirror had been distorted into a number of facets, each of which was contributing its own image. On inquiry we found that the dome had been open throughout the day while the workmen were busy with parts of the mounting, and it even seemed probable that the sun had shone, if not upon the mirror itself, at least upon the cover above it . . .
>
> Finally we decided to go back to the Monastery, but Hale and I made an engagement to meet at three o'clock in the morning at the telescope building. I doubt whether either of us slept in the interval, for we both arrived ahead of time. Jupiter was out of reach in the west, so we turned northward to the bright star Vega. With his first glimpse Hale's depression vanished: the mirror had resumed its normal figure during the long cool hours of the night, and the image of the star stood out in the eyepiece as a small sharp point of light, almost dazzling in its brilliancy. The success of this great instrument was fully assured.[25]

The 100 inch has always been known as the Hooker reflector after the

Fig. 7.4. The 100 inch Hooker reflector at Mt Wilson. Photograph courtesy of the Mt Wilson and Las Campanas Observatories.

Fig. 7.5. The dome of the 100 inch reflector at Mt Wilson. From a postcard presumably made in about 1920.

Fig. 7.6. Mt Wilson panorama in 1984. Photograph by the author.

donor of its mirror blank, but the lion's share of the funding for the construction of the rest of the telescope and the dome ($600 000) was borne by the Carnegie Institution. Like the 60 inch, the 100 inch was designed to be used as an f/5 Newtonian, an f/16 modified Cassegrain, or as an f/30 coudé. Since the mirror had 2.8 times the area of the 60 inch it could detect objects fainter than any ever photographed. Because of its light-gathering power and spectral resolution it could be used to study stellar spectra to a degree of detail never before achieved. As has been previously emphasized, having the largest optics is only important if one also has a solid mounting, excellent guiding, first-class auxiliary equipment (photographic emulsions, spectrographs, electronic detectors), an excellent observing site, and top notch personnel. All of these conditions were met (except for the electronic detectors, which came much later), and the 100 inch was soon making Mt Wilson *the* place for aspiring astronomers to come and work. Among the most outstanding achievements made with the Hooker reflector are those described below.

The direct measurement of stellar diameters[26]

Stars are so far away and the turbulence in our atmosphere is such that no amount of magnification can bring out disk-like images to the eye or on regular photographs.* In 1920 Albert Michelson and Francis Pease

---

* The method of speckle interferometry, developed by A. Labeyrie and his colleagues in the 1970s, is a newer method whereby the effects of the atmosphere are 'frozen' out by means of using short exposures (on the order of 1/100 of a second) and the speckles are reconstructed into a single image by means of computer processing (see Fig. 8.10 for an example).

mounted an *interferometer* at the top end of the 100 inch, and using the unsurpassed light-gathering power of the telescope and relying on its solid mounting and excellent guiding, they were able to measure the diameters of a number of stars in an indirect way. This was made possible by the moveable mirrors at the top end of the telescope, comprising the interferometer. Instead of giving stellar images at the eyepiece, one obtains interference fringes.[27] From the spacing of the mirrors, the stellar diameter can be obtained. For Betelgeuse (α Ori) they obtained a diameter of 0.047 arc seconds, which at the distance of the star (95 parsecs) corresponds to something larger than the size of the orbit of Mars about our Sun, almost half a billion miles. Other smaller or more distant stars would give correspondingly smaller angular diameters.

The determination of the true distances of what we now call galaxies[28]

On 1 January 1925 a paper by Edwin Hubble was read at a joint meeting of the American Association for the Advancement of Science and the American Astronomical Society in Washington, DC. Hubble had managed to discover more than a dozen Cepheid variable stars in each of the galaxies M 31 and M 33. Using Shapley's period-luminosity law, he was able to show that these stars, and consequently their parent nebulae, were situated at distances measured in hundreds of thousands of parsecs, thus much further than the extent of Shapley's Big Galaxy model. This achievement rendered our own piece of the universe smaller once again, for it implied that our own galaxy, large as it is, is just one in a sea of island universes.

The discovery of the expansion of the universe[29]

Following up the discovery by V. M. Slipher of Lowell Observatory that some galaxies are receding from us at large velocities, Edwin Hubble began a systematic study of the Doppler shifts in the spectra of galaxies. In 1929 he published a classic paper which demonstrated that the more distant galaxies were receding from us at a rate proportional to their distance. Hubble's law may be stated as follows:

$$V = H_o \times D,$$

where V is the velocity of recession (measured in km/s), D is the distance (usually measured in units of millions of parsecs, or megaparsecs), and H_o is the rate of expansion (measured in km/s/Mpc). Naturally, because of more local gravitational effects within galaxy clusters, individual galaxies can be receding from us more or less rapidly than the cluster average, and some of the nearest galaxies are actually moving towards us. Furthermore, the inaccuracies of determining the distances to galaxies lead to a rather wide scatter in a graph of velocity versus distance. Hubble's original estimate of the constant H_0 was about 500 km/s/Mpc. Modern research based on optical and radio data lead to a value of 50.3 ± 4.2 km/s/Mpc,[30] a reflection of better calibrations of distance.

However, the final answer is not in yet. In the late 1970s astronomers found that there is a strong correlation between the widths of the spectral lines of hydrogen observed in the radio region at a wavelength of 21 cm and the brightnesses of the galaxies observed in the infrared (notably the magnitudes at a wavelength of 1.6 microns), and that this relationship can be employed in a more self-consistent manner to more distant galaxies than can be done using radio and optical data. It is easy to understand why this is the case. The width of the hydrogen line is a direct measure of the rotational velocity of the hydrogen gas. This velocity is a measure of the amount of mass in the galaxy. Galaxies are pretty much the same as far as how their mass is used to make relative numbers of stars of different mass, and stars of a given mass and age have pretty much the same brightness. One can calibrate this system using nearby galaxies whose distances are known by means of the period-luminosity law for Cepheids or by means of the observation of novae. One can then measure the velocity widths for the hydrogen lines in more distant galaxies and also measure their near infrared magnitudes (which are less affected by the absorption due to interstellar dust than are optical magnitudes). One can then derive more accurate distances than can be done by optical methods, and from the Doppler shifts in the spectra (which are accurately measurable at any wavelength), one can derive the Hubble constant $H_0 = V/D$. This method has led to a value of 95 ± 4 km/s/Mpc,[31] significantly different than the more traditional, optically derived value.[32]

The exact value of the Hubble constant is extremely important, for it specifies the age and size of the universe. Since the speed of recession can not be greater than the speed of light, it follows that the limit of the visible universe is at a distance such that the galaxies there are receding at the speed of light. In other words, at what distance will $H_0 \times D =$ speed of light?

That distance will be:

$$D = \frac{\text{speed of light}}{H_0} = \frac{300\,000 \text{ km/s}}{\left(\dfrac{H_0}{100}\right)100 \text{ km/s/Mpc}} = \frac{3000 \text{ Mpc}}{\left(\dfrac{H_0}{100}\right)}.$$

If $H_0 = 100$ km/s/Mpc, the size of the universe will be 3000 Mpc, or 9.8 billion light years. The light from those galaxies receding at the velocity of light will have been travelling towards us for almost 10 billion years, so the universe will be at least 10 billion years old. But if $H_0 = 50$ km/s/Mpc, the most distance galaxies would be 6000 Mpc away and the universe would be at least 20 billion years old.

A check on this can be found in the determination of the ages of the oldest stars in our galaxy. For, if there are stars 18 billion years old in our galaxy, how can the universe, of which they are a subset, be younger? The oldest stars in our galaxy are to be found in globular clusters. Allan Sandage has found that the globular clusters M 92 and M 15 are 16–20 billion years old.[33] This would be strong evidence in favor of the lower value of the Hubble constant ($H_0 = 50$), but it must be said that the astronomy of very distant objects is more of a qualitative, rather than a quantitative, science, and there is more work to be done on determining the Hubble constant.

Since the universe is expanding, that obviously means that in the past it was smaller. If the change of size can be reckoned back in time to the ultimate consequence, it means that the universe originated in a great primeval fireball, the Big Bang.

Will the universe expand forever? That too depends on the Hubble constant. Since all matter in the universe exerts a gravitational force on all other matter, it follows that there is a critical density of the universe that will lead to a halt of the general expansion.[34] The critical density is

$$\rho_c = 3\,H_0^2/8\pi G,$$

where G is Newton's gravitational constant. This can be converted to the following equation:

$$\rho_c = 1.88 \times 10^{-29}(H_0/100)^2 \text{g/cm}^3.$$

Whether the Hubble constant is equal to 50 or 100 km/s/Mpc, the critical density is the equivalent of several hydrogen atoms per cubic meter, not much you might say, but for now astronomers have only been able to account for about 10% of the critical density. The universe may very well expand forever. This strikes many as philosophically unpalatable, but the opposite scenario, in which the universe crashes back on itself, vaporizing everything in a fiery Big Crunch, is not too pleasant to consider for the long-term future either.[35]

The recognition of the two principal populations of stars[36]

During the Second World War, when the night lights of Los Angeles were blacked out, Walter Baade pushed the 100 inch to its limits and resolved the nucleus of the Andromeda Nebula and its companions M 32 and NGC 205 into individual stars. He found that while the most luminous stars in the

Fig. 7.7. Las Campanas Observatory in the Chilean Andes. From left to right, the domes of the 1.0 m reflector, the University of Toronto 24 inch, and the 2.5 m du Pont reflector. Photograph courtesy of the Mt Wilson and Las Campanas Observatories.

spiral arms of the Andromeda Nebula were hot, blue O and B stars, the brightest stars in the nucleus and in the companion galaxies were red stars much like the brightest members of globular clusters in our own galaxy. The hot, young stars are known as Population I stars, while the older, cooler stars that occupy the nuclear regions (and halos) of spiral galaxies are Population II. Elliptical galaxies are made up of Pop II stars.

In 1952 Baade announced that Cepheid variables of Pop I are 1.5 magnitudes brighter than Cepheids of Pop II. This was one of the first discoveries made on the basis of photographs taken with the 200 inch telescope. The bottom line of this discovery of the two populations of stars and the recalibration of the zero point of the period-luminosity law meant that all galaxies were really twice as distant as previously thought. In one fell swoop Baade cut the accepted value of the Hubble constant in half and doubled the size and age of the universe.

The Two-Micron Sky Survey

One of the projects carried out at Mt Wilson that must be mentioned is the *Two-Micron Sky Survey* (also known as the Infrared Catalogue, or IRC), carried out from 1965 to 1968 under the direction of Gerry Neugebauer and Robert Leighton and involving many Caltech undergraduate and graduate

Fig. 7.8. The Irénée du Pont 2.5 m reflector at Las Campanas. Photograph courtesy of the Mt Wilson and Las Campanas Observatories.

students. Using a 62 inch telescope with an f/1 aluminized epoxy mirror, the sky from declination $-33°$ to $81°$ was scanned at wavelengths of 0.84 and 2.2 microns. A total of 20 000 sources was found, of which 5612 were brighter than magnitude 3.0 at a wavelength of 2.2 microns (equivalent to a flux of 40 Janskys). This was the first systematic survey of the sky at infrared

Fig. 7.9. West coast US night sky brightness, which has seriously affected observing conditions on Mt Wilson and which threatens to affect seriously the observing programs at Palomar Mountain. Produced from USAF DMSP (Defense Meteorological Satellite Program) film transparencies archived for NOAA/NESDIS at the University of Colorado, CIRES/National Snow and Ice Data Center.

wavelengths. Some sources, such as IRC + 10216, are still referred to by their designations in this catalogue.

Since 1976 the astronomers of Mt Wilson have operated a southern hemisphere observatory station at the 8235 foot summit of Cerro Las Campanas, 120 miles northeast of La Serena, Chile. The establishment of this station was the culmination of Hale's original plans of 1902 to have a permanent southern station for the Carnegie Institution. Hale, who died in 1938, did not live to see its inception, but he would have been proud of the results.

The Mount Wilson and Las Campanas Observatories (as they have been known since 1980) operate the Irénée du Pont 2.5 m telescope at Las Campanas, along with a smaller, 1.0 m reflector. They are naturally outfitted with modern electronics and computer controls. Astronomers at Las Campanas enjoy extremely dark skies, and in a good year there might be up to 300 useable nights. The University of Toronto also operates a 24 inch telescope there.[37]

The term *extremely dark skies* is important here, for the skies at Mt Wilson are no longer dark, even at new Moon (see Fig. 7.9). The funding for astronomy goes through various cycles, and in spite of various grand plans that astronomers have, the reality of the mid 1980s is such that there is not enough money to go around necessarily to keep even such a famous institution as Mt Wilson open. In the summer of 1984 the Carnegie Trust decided that the 100 inch at Mt Wilson would be shut down, effective 1 July 1985. The 60 inch reflector and the solar telescopes would be operated for another year. The observatory 'top brass' has emphasized that their funds would be better spent supporting the dark sky site in Chile, because it is better for extragalactic studies, their principal interest.

Many large telescopes are planned for the near future by astronomers around the world. Two astronomers recently asked:

> Will the coming generation of ultra-large telescopes lead to the closure of more and more of the smaller but still scientifically productive instruments? Or does the Hooker telescope have yet more glorious years ahead of it? We suggest that astronomy is neither so rich that it can afford to close Mt Wilson, nor so poor that it is compelled to do so.[38]

When W. J. Hussey of Lick Observatory had done site testing for the Carnegie Institution at the turn of the century for the establishment of a solar observatory and possibly also a southern hemisphere observatory, he eventually recommended Mt Wilson for the solar observatory, but he had had a chance to test the seeing at Palomar Mountain, a 5600 foot peak northeast of San Diego, California. He described it as a 'hanging garden above the arid lands'.[39] It was passed over as an observatory site in 1903 because of its general inaccessibility, but later, when a real road was paved to the summit of Palomar Mountain, its remoteness became an advantage. The sky was darker and would likely remain so.

Once the 100 inch was working Hale dreamt of a still larger telescope. Francis Pease made plans for a telescope as large as 300 inches, but Hale and others felt that that would be too big a step. They hoped that a 200 inch telescope could be built and started out to find funding for it. In a 1928 article, quoted at the beginning of this chapter, Hale talked of the possibilities of large telescopes. A copy of this article in proof was sent to Wickliffe Rose at the Rockefeller Foundation. The end result was that the International Education Board of the Rockefeller Foundation would provide $6 000 000 for the construction of a 200 inch telescope. It would be built with the full cooperation of the Mt Wilson staff, but the telescope upon completion would become part of the California Institute of Technology. It was Rose who strongly pushed for the new telescope to be run by an educational institution rather than a research organization such as the Carnegie Institution.

The first question was – could a mirror 200 inches in diameter be cast? That depended on the material to be used. Plate glass was out of the question on two counts: the Saint-Gobain glassworks, which had cast the Mt Wilson 60 inch and 100 inch mirrors, had been destroyed during the First World War; and better materials were available by that time. The aim is to use the material which is affected least by temperature. In other words, the ideal material would have a very low coefficient of expansion. The best material known was fused quartz.

The General Electric Company was contracted to make some tests, and while this provided quartz mirror blanks up to 60 inches in diameter, the tests greatly overran the cost and time estimates. It was concluded that a different material would have to be used. It was decided to use borosilicate glass, commonly known as Pyrex, which normally has a coefficient of expansion five times that of fused quartz, but one-third that of plate glass. By increasing the quartz content of the Pyrex mix, one can make Pyrex with a coefficient of expansion three times that of fused quartz.[40] Even so, a solid Pyrex disk 200 inches in diameter would take a year to cool, so the mirror was designed to have a ribbed back. This would greatly reduce the cooling time and also the mass of the finished mirror, but it also necessitated a special mirror mounting to compensate for sagging of the mirror while the telescope was set at different sky positions.

The Mt Wilson staff and the Corning Glass Works, who would cast the Pyrex blank, decided to work their way up to a 200 inch piece of glass by

Fig. 7.10. The dome of the 200 inch Hale reflector at Palomar Observatory.
Photograph by the author.

casting disks of 30, 60 and 120 inches. (The 120 inch, which was used for testing the 200 inch, eventually became the main mirror of the Lick Observatory Shane reflector.) A successful 200 inch blank was cast in December of 1934, which made its way to Pasadena by special train in the spring of 1936. The mirror was ground to an f/3.3 figure and only finished while installed in the telescope in November of 1947.

The details of the engineering aspects of the Palomar giant have been described many times.[41,42] It was the first telescope which allowed an observer to ride along at the prime focus, on the inside of the 72 inch diameter prime focus cage. The telescope can also be operated at the f/16 Cassegrain focus or at the f/30 coudé focus. The mounting has a large horseshoe-shaped bearing at the north end, while the smaller southern bearing is spherical. The 55 foot tube assembly weighs 530 tons, and the north and south bearings float on a thin film of Mobil Flying Horse telescope oil[43] at a pressure of 300 pounds per square inch. The 1000 ton dome is 135 feet high and 137 feet in diameter.

To appreciate the size of the Palomar telescope one may consider that the Yerkes telescope, once the largest in the world, could fit neatly inside the Cassegrain hole of the 200 inch mirror. What was considered large at the beginning of the century was now regarded as an acceptably disposable part of the new telescope. The numbers I have just quoted mean that a 13 story

Fig. 7.11. The 200 inch Hale reflector. Palomar Observatory photograph.

building could fit inside the dome, either standing up or lying down. The mirror has the power of several hundred million human eyes. If one looked through an eyepiece at the focus of the 200 inch one could see objects much fainter than 20th magnitude, and photography or modern electronic detectors could extend that further by a few magnitudes.

The 200 inch was created to supplement rather than duplicate the performance of the 100 inch. The 200 inch was *not* a scaled-up version of the 100 inch, which, as we have mentioned, was a scaled-up version of the 60 inch. Originally, there was some thought that the 200 inch might be situated in the southern hemisphere, but such a large telescope would have a very small field of view, so it would be best utilized for the detailed study of known objects or yet to be understood objects in the northern sky, whose study would be greatly enhanced by the unsurpassed light gathering power of the 200 inch. Besides, the mounting of the 100 inch does not allow observations with that telescope further north than $+64°$ declination. This north polar cap of the sky would be accessible to the 200 inch, as its mounting allowed the huge newcomer to be slewed all the way to the north celestial pole.

There are three other telescopes at Palomar Mountain, the 48 inch

Schmidt telescope, an 18 inch Schmidt, and a more standard 60 inch reflector. A Schmidt telescope, or Schmidt camera, is a special photographic telescope which utilizes a spherical, rather than parabolic, main mirror and a smaller correcting lens placed at the front end. The focal plane is in the middle of the tube, halfway between the correcting lens and mirror. Furthermore, the focal plane is not flat, so the photographic plates must be bent for all the stars to be in focus. Why go through so much trouble? To eliminate the effect of coma associated with reflectors (which gives the comet-like images at the edge of the field) and to achieve a very large field of view. The 48 inch Schmidt, which has a 72 inch main mirror and a 48 inch correcting lens, was used for the National Geographic Society-Palomar Observatory Sky Survey carried out in the 1950s. Each plate is 6° on a side and reaches objects fainter than 20th magnitude.

While the Schmidt cameras have been used in survey work, looking for what might be out there at a given epoch, the 200 inch has been used for the detailed study of already known objects. By 'known' I mean only that something peculiar was noted at a particular set of celestial coordinates and the great light-gathering power of the 200 inch was brought to bear on the problem. No better example could be given than the discovery of quasars in 1962–3.[44] First, on the basis of an occultation of the radio source 3C 273 by the Moon the object's position was pinned down accurately enough to associate it with a 13th magnitude star-like object. But when Maarten Schmidt of Palomar obtained a spectrum of this object with the 200 inch, he was at first puzzled by the wavelengths of the emission lines. The only explanation was that we were observing spectral lines which were red-shifted 16%. If the object were at a distance in accord with Hubble's law, its speed of recession of 16% the speed of light would place it hundreds of megaparsecs away.

Since this discovery almost 25 years ago, astronomers have expended a great deal of telescope time nailing down the positions of these quasi-stellar radio sources, measuring their red shifts, and arguing about their nature. The best evidence obtained so far indicates that they are at their cosmological distances (in accord with Hubble's law), which is to say that their redshifted spectral lines are so shifted because the objects are actually receding from us at tremendous velocities. They are very distant and also very luminous, but not very large compared to galaxies. In fact, the best evidence is that they are active nuclei of galaxies which are forming. The largest redshift obtained so far has a wavelength shift of $\Delta\lambda/\lambda = 4.11$,[45] indicating a recessional speed of 93% the speed of light[46] and a distance of 12 billion light years (assuming $H_0 = 75$ km/s/Mpc). It must be added that all astronomers are not in agreement with the view of quasars I have just given. Some argue that another mechanism gives rise to the redshifted spectral lines. They further argue that quasars are nearby objects (as opposed to being at the far reaches of the universe), which correspondingly decreases their derived rates of energy production.

The Palomar telescope has naturally been used as the big gun of American observational astronomy for the past 40 years.[47] Of the great number of key projects carried out on it are: Baade's work on the different populations of stars, which we have already mentioned; Allan Sandage's determination of the modern value of the Hubble constant, the age of the universe, and his correspondingly consistent value of the ages of globular clusters. We must also mention Jesse Greenstein's definitive studies of the nature of white dwarf stars, Becklin and Neugebauer's pioneering infrared studies, and observations of supernovae and supernova remnants by Baade, F. Zwicky and R. Minkowski.[48] There has been a host of projects in galactic and extragalactic astronomy: the classification of galaxies, the variation of the abundances of elements within galaxies, studies of galactic dynamics for individual galaxies and clusters of galaxies, and studies of the dynamics of the stars in globular clusters on the basis of very accurate radial velocities obtained with a device called a radial velocity spectrometer. Unfortunately, constraints on time and space prevent me from doing better justice to this discussion.

One of the most significant changes in the use of the 200 inch over the years has been the type of light detector used. At the start a great deal of work was done photographically at prime focus. Now most work is done at the Cassegrain focus with detectors such as charge-coupled devices (CCDs), which are solid state detectors capable of very high quantum efficiency.*

A good example of the use of CCD technology is the recovery of Comet Halley on 16 October 1982 by David Jewitt and G. Edward Danielson using the Palomar 200 inch and a CCD camera.[49] At that time Comet Halley was at magnitude 24.2. It was beyond the orbit of Saturn. It was 1212 days before scheduled perihelion passage on 9 February 1986. This is a tremendous leap beyond the astronomical capabilities at the turn of the last century. By

---

* If we consider an analogy of carrying water from a stream to the house in a bucket, an efficient means of doing it would ensure that the least amount of water were spilled. Photographic plates have about 1% efficiency. For a photomultiplier tube (*c.* 1940s technology) the quantum efficiency might be a few percent, meaning that 30 times as much light is spilled as gets to be used. But the modern CCDs have up to 90% quantum efficiency, thus allowing much fainter objects to be detected. Also, the output of a CCD can be directly read into a computer for automatic analysis.

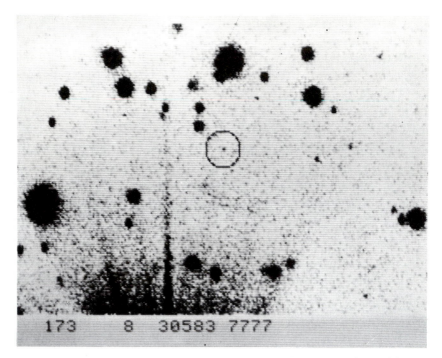

Fig. 7.12. CCD image of Comet Halley on 16 October 1982, obtained by G. Edward Danielson and David C. Jewitt with the 200-inch Hale reflector. North is to the left. California Institute of Technology photograph.

comparison, for the 1910 passage of the comet it was only recovered 221 days before perihelion. By the time the Space Telescope is in orbit, we will have the capability of following Comet Halley all the way out to the far end of its orbit beyond Neptune. That will be 37 years before the next perihelion passage, a 10-fold improvement on the 200 inch, but by a 94 inch telescope!

The Palomar telescope reigned as the world's largest until the Soviets completed their 6 meter altazimuth telescope in 1976.[50] However, a telescope today means telescope-plus-detectors-plus-sky, and there is no doubt that the staffs of the Mt Wilson and Palomar Observatories still have the upper hand on the Soviets when it comes to big telescope usage. But these are not the only large telescopes in the business. Within the last 15 years a number of 3–4 meter class optical and infrared telescopes have appeared on mountaintops in Arizona, Hawaii, Chile, Australia and Spain. The concept of the 'astronomical capital of the world' breaks down because it depends on the wavelength one is working at and whether one is studying objects in the southern sky or northern sky. Since the Second World War we have witnessed the birth of radio astronomy, which does not require dark mountain sites. The space age has brought us space age astronomy satellites which operate at X-ray, ultraviolet, and infrared wavelengths. Astronomical research has been pioneered within the cores of supercomputers. The flavor of astronomy is changing greatly as we head toward the twenty-first century. Let us turn to a discussion of these developments.

8

The present

The southern hemisphere

So far our discussion has rightly centered on astronomical institutions in the northern hemisphere, for that is where the greatest advances in technology were made, that is where most of the people were, and that is where most of the land on Earth is. It has been said that the northern hemisphere contains astronomers, while the southern hemisphere contains stars, as anyone who has seen the southern sky will testify.

Owing to the sphericity of the Earth, European and American astronomers could not see the far southern sky. For an observer at 40° north latitude, the north celestial pole is always 40° above the horizon, and the south celestial pole is 40° below the horizon. If such an observer had a clear southern horizon, he could theoretically see a star at 50° south declination for its brief moment above the horizon but, in practice, stars further south than 25° south declination went unstudied until relatively recent times.

The first systematic survey of the southern skies by an astronomer was made by Edmond Halley (1656–1743), eventually the second Astronomer Royal. At the age of 20 he sailed to the island of St Helena (16° south latitude) and determined the positions of 341 southern stars. For this he was referred to by Flamsteed as the 'Southern Tycho'.[1] Seventy five years later the Frenchman Nicolas de Lacaille (1713–62) went to South Africa, where from 1751 to 1753 he carried out a variety of astronomical projects, the most impressive being the cataloguing of the positions of 9766 stars between 23° south declination and the south celestial pole.[2,3] For this he used a 1/2 inch telescope of 28 inches focal length! It was Lacaille who named many of the southern constellations such as Telescopium, Microscopium and Antlia (the air pump), a change of pace from the mythological orientation of the past. Lacaille's measurements of the shape of the Earth have already been mentioned (p. 69).

Another three generations would elapse before the appearance of the next great southern observer, once again a temporary visitor from the northern hemisphere: John Herschel, the son of William Herschel. In 1834 John Herschel arrived at the Cape of Good Hope with his wife and children and an $18\frac{1}{4}$ inch reflecting telescope of 20 foot focal length (including two spare speculum metal mirrors). Over the next 4 years he catalogued 1707 nebulae and clusters and listed 2102 pairs of binary stars. He also made star gauges (the number of stars per unit area of the sky) as had his father for the northern sky, the family thus providing whole sky coverage of the general structure of the Milky Way.[4] He also observed the 1835–6 apparition of Halley's Comet.

Before John Herschel had set foot in South Africa, the first permanent observatory had already been founded there. In 1820 the British Board of Longitude, the supervising body of Greenwich Observatory, had resolved to establish the Royal Observatory at the Cape of Good Hope.[3,5–9] The first Astronomer Royal at the Cape (1820–31) was the Rev. Fearon Fallows, who produced a catalogue of 273 southern stars but published little else. He spent his time primarily arranging for the construction of the permanent observatory building (1825–7) and arranging for the installation of the instruments (by 1828), including a 6 foot mural circle and a 10 foot transit instrument. Fallows died in 1831 from scarlet fever and was briefly (1831–3) replaced by Thomas Henderson, who, with his assistant Lt. William Meadows, made 10 000 measurements of star positions in 1832. After Henderson became the first Astronomer Royal for Scotland in 1834, he got around to reducing his South African observations of α Centauri and in 1838 became one of the first three astronomers to publish a bona fide trigonometric stellar parallax (the others being W. Struve and W. Bessel).

The third Astronomer Royal at the Cape (1833–70) was Thomas Maclear, whose remeasurement of the Lacaille meridian arc has already been mentioned (p. 69). Under Maclear's direction positional astronomical work greatly increased in scope. From 1837 to 1850 the Bradley zenith sector (with which Bradley had discovered aberration) was on loan to Maclear. He used it for accurate latitude determinations as part of the meridian arc survey. A duplicate of the Airy transit circle arrived in South Africa in 1854, which allowed for the determination of star positions more readily than with the older method of using a transit instrument and a mural circle.

But while Maclear and his assistants accumulated a large number of measurements, they did little in the way of data reduction, a task which fell to his successor (1870–9), Edward J. Stone. George Airy set as Stone's main task the clearing up of the backlog of unreduced observations, a task he did finish as Astronomer Royal at the Cape. A catalogue of 12 441 stars was published in 1880. However, this task precluded the taking of new observations, so that by the time Stone left, the instruments and the building itself were in a generally run down state.

The most significant Astronomer Royal at the Cape was David Gill, who lived from 1843 to 1914 and was Astronomer Royal there from 1879 to 1907. Under Gill the observatory embarked on a new direction based on the development of photographic methods. Gill's photograph of the comet of 1882 demonstrated that photography could also be used for making photographs of star fields, and hence star catalogues. The end result was the *Cape Photographic Durchmusterung*, eventually published in 1896–1900 and containing the positions and magnitudes of 454 875 stars south of 18° south declination. J. C. Kapteyn supervised the reduction of the plates in The Netherlands with the assistance of local convicts.[10] Gill was also one of the prime movers behind the *Carte du Ciel* project (see pp. 72–3).

Gill made extensive use of the heliometer, an instrument with the prime objective sawed in half, the relative positions of which are adjusted by means of a micrometer screw. It allows accurate determinations of the relative positions of stars and other point-like objects. The Cape astronomers used heliometers for the determination of trigonometric parallaxes of stars and the determination of the solar parallax using systematic observations of minor planets against the background of stars. Gill's value of 8″.80 for the solar parallax (corresponding to an Earth–Sun distance of 92.9 million miles) was used until the 1960s.

The Astronomers Royal at the Cape during the twentieth century were S. S. Hough (1907–23), Harold Spencer Jones (1923–33; subsequently Astronomer Royal at Greenwich, 1933–55), John Jackson (1933–50), and R. H. Stoy (1950–68). Under Gill and these men various instruments were added to the Cape Observatory's arsenal: the 13 inch *Carte du Ciel* photographic refractor (1890); a 24 inch photographic refractor coupled with an 18 inch visual refractor on the same mounting (1901); a reversible transit circle (1901); an 18 inch reflector (1954); a 30 inch reflector (1961); and a 40 inch reflector (1963). The reflectors were used for more astrophysically oriented projects: photoelectric photometry and spectroscopy of stars. For the most part the Royal Observatory at the Cape carried out positional work such as the determination of 40 000 proper motions from second epoch plates taken after a span of 25 years with the astrographic refractor, the determination of stellar parallaxes, and further determinations of fundamental stellar positions from meridian observations.

Solar astronomy was also carried out. As early as 1875 a photoheliograph was being used for solar photographs. Regular photographs were taken for many years beginning in 1910, and in some years 350 days were clear enough for solar photography. Few places on Earth could provide such consistently clear sky. Better time resolution in solar photography (many pictures taken each day at a rate of one per minute) was achieved by means of a Lyot heliograph installed in 1958.

The official life of the Royal Observatory at the Cape came to an end in 1972 when it became part of the South African Astronomical Observatory (SAAO).[11] The base remained at the Cape, but the two largest reflectors were moved to Sutherland, 230 miles from Cape Town. The other two constituents of the SAAO are briefly described below.

A meteorological station and time service was established outside Johannesberg in 1903, and at Gill's suggestion the Cape assistant R. T. A. Innes was appointed director. This was the Transvaal Observatory, which became the Union Observatory in 1912 after the Union of South Africa was formed. Innes, who had previously been a well-to-do wine merchant in Australia prior to his career in astronomy, was a keen double star observer. He discovered the closest star (other than the Sun), Proxima Centauri. As a result of his efforts, along with the influence of Gill, Kapteyn, and the Pulkovo Observatory director Backlund, Union Observatory obtained a $26\frac{1}{2}$ inch refractor. It went into operation in 1925. Double star work was a prime area of activity at the Union Observatory under Innes and the third and fourth directors, W. H. van den Bos and W. S. Finsen. By 1970 more than 4000 double stars had been discovered at the observatory, which in 1961 had been renamed the Republic Observatory. In 1972 it became part of the SAAO. The $26\frac{1}{2}$ inch refractor still resides in Johannesberg, but a 0.5 m f/18 Cassegrain reflector from the former Republic Observatory was moved to Sutherland.

The third constituent of the SAAO is the Radcliffe Observatory, originally founded in 1772 in Oxford. Under the direction of H. Knox Shaw it moved, more in name than in body, from England to Pretoria, South Africa, in 1938. The old instruments were left behind. The Radcliffe 74 inch reflector became operational in 1948. It was principally designated for determining stellar radial velocities in order to complement programs in the northern hemisphere, notably the radial velocity work carried out with the 72 inch reflector of the Dominion Astrophysical Observatory (British Columbia, Canada; operational 1918). Along with the twin 74 inch of the Mt Stromlo Observatory in Australia (operational in 1953), the Radcliffe telescope was the largest in the southern hemisphere until 1974. After the establishment of the SAAO in 1972, the 74 inch was moved from Pretoria to Sutherland.

There have been a number of other observing stations established in South Africa. As already mentioned (Chapter 6, note 25), the Harvard Boyden station was moved from Peru to Bloemfontein in 1927. The University of Michigan also established a station in South Africa that year. This was the Lamont–Hussey Observatory, containing a 27 inch visual refractor for double star work. The Michigan station operated for over 25 years. Yale Observatory operated a station on the grounds of the University of Witwatersrand from 1925 to 1952. The principal instrument there was a 26 inch photographic refractor used for determining stellar parallaxes. In 1952 the equipment was moved to Australia. Last, but not least, Leiden Observatory installed a twin 16 inch photographic refractor at the Union Observatory in 1938. After the Second World War, it was moved to the Union Observatory's annex at Hartbeesport, where the Leiden astronomers also installed a 36 inch reflector for making photoelectric measurements.

From about 1830 to 1970 the astronomical capital of the southern hemisphere was clearly South Africa. There arose a large number of survey instruments dedicated to measuring star positions, proper motions, parallaxes, magnitudes, radial velocities and discovering variable stars. For the most part positional astronomy was carried out before the Second World War, but with the establishment of reflectors, more photometry and spectroscopy was done. Well over a dozen professional telescopes in the 0.5–1.9 m range were spread out all over the country, a testament to the generally excellent seeing conditions available and to the opportunities of survey work of the southern sky, which incidentally contains the Magellanic Clouds and the center of our galaxy.

The astronomical capitals of the southern hemisphere today are Chile and Australia. Chile has three major observatories (European Southern Observatory, Cerro Tololo Inter-American Observatory, and Las Campanas), sporting many telescopes both large and small (see Table 8.1, Figs. 7.7, 7.8 and Figs. 8.1–8.3). That takes care of a lot of southern hemisphere optical astronomy. The Australians were pioneers in radio astronomy (see later on in this chapter), and the Anglo-Australian Telescope (AAT; Fig. 8.4) is in a class by itself when it comes to southern hemisphere infrared astronomy. The site of the AAT, Siding Spring Mountain (Fig. 8.5), is also the site of the United Kingdom Schmidt Telescope, which completed the Sky Survey begun at Palomar, but to a significantly fainter limiting magnitude, and the site of the Australian National University's 2.3 m New Technology Telescope which was dedicated in 1984.

Table 8.1. *The world's largest ground-based optical and infrared telescopes*[a]

Size Ranking	

European

1   Bol'shoĭ Teleskop Azimutal'nyĭ
           Special Astrophysical Observatory
           Zelenčhukskaĭa, Russian Soviet Federal Socialist Republic
           Mt Pastukhov (2100 m/6900′)
           6.0 m   1976
           (*ST*, Nov. 77, pp. 356–62)

4   William Herschel Telescope
           Observatorio Roque de los Muchachos
           La Palma, Canary Islands, Spain (2332 m/7650′)
           4.2 m 1987
           (*Vistas in Astronomy*, **28**, 1985, pp. 531–53)

11   German–Spanish Astronomical Center
           Calar Alto, Spain (2160 m/7100′)
           3.5 m   1985
           (*ST*, Apr. 1980, pp. 279–86)

15 (tie)   Shajn Telescope
           Crimean Astrophysical Observatory
           Simeis, Ukrainian Soviet Socialist Republic
           2.6 m/102″   1961

15 (tie)   Byurakan Astrophysical Observatory
           Armenian Soviet Socialist Republic
           Mt Aragatz (1500 m/5000′)
           2.6 m/102″   1976
           (*ST*, Apr. 1971, pp. 217–19; Jan. 1977, p. 13)

18   Isaac Newton Telescope
           Observatorio Roque de los Muchachos
           La Palma, Canary Islands, Spain (2336 m/7660′)
           2.54 m/100″   1984
           (*J. Hist. Astr.*, **13**, 1982, pp. 1–18; *Vistas in Astronomy*, **28**, 1985,
           pp. 483–94)

Southern Hemisphere

5   Cerro Tololo Inter-American Observatory
           Cerro Tololo, Chile (2160 m/7100′)
           4.0 m   1976
           (*ST*, Feb. 1968, pp. 72–6; Jan. 1974, pp. 11–16)

Table 8.1. (*Contd.*)

Size/Ranking	

6   Anglo-Australian Telescope
               Anglo-Australian Observatory
               Siding Spring Mountain, Australia (1165 m/3820′)
               3.9 m   1974
               (*ST* Oct. 1975, pp. 225–8; *Quart. J. Royal Astr. Soc.*, **26**, Dec. 1985, pp. 393–455)

10   European Southern Observatory
               Cerro La Silla, Chile (2400 m/7850′)
               3.6 m   1976
               (*ST*, Jan. 1974, pp. 11–16; Feb. 1977, pp. 97–103)

17   Irénée du Pont Telescope
               Mount Wilson and Las Campanas Observatories
               Cerro Las Campanas, Chile (2510 m/8235′)
               2.57 m/101″   1977
               (*ST*, Jan. 1974, pp. 11–16)

United States

2   George Ellery Hale Telescope
               Palomar Observatory
               Palomar Mountain, California (1700 m/5600′)
               5.08 m/200″   1948

3   Multiple Mirror Telescope[b]
               Whipple Observatory
               Mt Hopkins, Arizona (2606 m/8550′)
               (*ST*, Jul. 1976, pp. 14–21; Jul. 1979, pp. 23–4)

7   Nicholas U. Mayall Telescope
               Kitt Peak National Observatory
               Kitt Peak, Arizona (2100 m/6900′)
               3.8 m   1973
               (*ST*, Jan. 1973, pp. 10–17)

8   United Kingdom Infrared Telescope
               Mauna Kea, Hawaii (4200 m/13 780′)
               3.8 m   1979
               (*ST*, Jul. 1978, pp. 22–4)

9   Canada–France–Hawaii Telescope
               Mauna Kea, Hawaii (4180 m/13 720′)
               3.6 m   1979
               (*ST*, Apr. 1977, pp. 254–6)

Table 8.1. (*Contd.*)

Size Ranking	
12	C. Donald Shane Telescope
	Lick Observatory
	Mt Hamilton, California (1277 m/4190′)
	3.05 m/120″ 1959
13	NASA Infrared Telescope Facility
	Mauna Kea, Hawaii (4160 m/13 650′)
	3.0 m 1979
	(*ST*, Jul. 1978, pp. 25–7; Dec. 1980, pp. 462–5)
14	McDonald Observatory
	Mt Locke, Texas (2070 m/6791′)
	2.7 m/107″ 1968
	(*ST*, Dec. 1968, pp. 360–7)

[a]Included are telescopes with effective apertures greater than or equal to 100 inches (except for the Mt Wilson 100 inch, which was shut down in 1985). Given are the telescope's ranking in terms of light-gathering power, the official telescope name (if it has one), the observatory name, the site, the elevation of the site (probably good to ± 30 m/100 feet), the size of the main objective, and the year the telescope went into operation. Since most of these telescopes are relatively new, references are also given to articles describing the observatory or telescope (*Sky and Telescope* magazine = ST). See also *ST*, Apr. 1981, pp. 303–7, *Mercury*, Sep.–Oct. 1982, pp. 142–3, and these books: Kirby-Smith, H. T., *US Observatories: a directory and travel guide*, New York: Van Nostrand, 1976; Marx, Siegfried and Pfau, Werner, *Observatories of the World*, New York: Van Nostrand, 1982.
[b]The MMT has six 72 inch mirrors. The light-gathering power is equivalent to a 4.5 meter telescope.

The growth of modern astronomy

The evaluation of astronomical research and astronomical institutions of the present era is fraught with ideological dangers, for it is impossible to be objective. Many astronomers whose institutions I would discuss are still active; a statement as to where the 'best' astronomy is being done would be biased and unfair, and I would not want to incur the wrath of all my colleagues. I do not think, however, that the following thoughts are wrong: as it happens with institutions, one place may have a hard time getting on its feet, so there arise various stories as to how 'they' went wrong along the way; but observatories are not regular businesses, and even one that gets off to a

Fig. 8.1. European Southern Observatory, La Silla, Chile. Photograph courtesy
of ESO.

slow start eventually gets its act together. The funding agencies see to that.
On the other hand, there are institutions that once had many great
successes, but due to administrative problems, funding constraints, or
deteriorating viewing conditions (largely stemming from urban growth),
some great observatories are not as great as they once were in the cosmic
scheme of things.

Fig. 8.2. Cerro Tololo Inter-American Observatory, Chile. Photograph from the National Optical Astronomy Observatories.

Depending on one's individual research interests and what one considers an acceptable place to *live and work*, astronomers have widely differing views about the best place to be. For ground-based optical and infrared astronomy, the observatories referred to in Table 8.1 and pictured in Chapters 7 and 8 would all be extremely desirable places to work, and the results obtained on the telescopes sited there would constitute a large percentage of the 'world-beating' results of modern optical and infrared observational astronomy. Naturally, light-gathering power is a vital parameter for doing the research, and one would expect that the most advanced equipment is being used by the best-qualified astronomers. However, this does not imply that one necessarily needs a large telescope to produce useful scientific work. It turns out that while the larger telescopes produce more scientific papers *per telescope*, the smaller telescopes (0.4–1.0 m range) actually produce more results *per research dollar* than the larger ones.[12] It must be stressed to amateur astronomers and professionals alike that good astronomy research is not done only on the faintest objects observed with the largest (and most expensive) telescopes.

Let us now discuss some aspects of the growth of modern astronomy: the

Fig. 8.3. The 4 m reflector at Cerro Tololo. Its twin is the Nicholas U. Mayall Telescope at Kitt Peak. Photograph from the National Optical Astronomy Observatories.

number of observatories, the number of telescopes, the number of astronomers, and the productivity patterns of the astronomers. I allow myself the 'objectivity' of a statistical treatment, thus hopefully saving me from the ire of astronomers and observatory directors not mentioned, or not mentioned to their liking.

**Fig. 8.4. The Anglo-Australian Telescope, Siding Spring Mountain, Australia.
Photograph from the Royal Observatory, Edinburgh.**

In a study of the establishment of observatories over the years 1560 to 1930 Dieter Herrmann has shown that the number of observatories has increased according to an exponential law of growth, with the number of observatories (both private and institutionalized) doubling every 35 years or so. If we restrict ourselves to astronomical institutions that are not private and not primarily dedicated to educational purposes, the doubling time is

Fig. 8.5. Siding Spring Observatory, Australia. Photograph from the Royal Observatory, Edinburgh.

about 45 years, but the same effect still holds: with an increase of national economic power, an increase in the number of people and, correspondingly, the number interested in astronomy, there will be an increase in the number of observatories. By 1930 some 660 observatories (both private and institutional) could be enumerated.[13] The 1986 edition of the *Astronomical Almanac* lists 378 optical observatories and 89 radio observatories. Not surprisingly, many observatories once active have closed, and a complete list of professional observatories is difficult to compile. Almost certainly, less than 50% of all observatories ever founded are still active. It has required more and more significant outlays of money to push back the outside of the envelope of astronomical inquiry. The progress of observational astronomy has required the establishment of remote mountain observatories, space satellites and interplanetary probes. There is no reason to suspect that the exponential growth of observatories stopped in 1930.

Today it is not so much 'How many observatories are there?' but, 'How many telescopes are there and where are they located?' The modern era has seen the diminishing popularity of the single-new-telescope-observatory in

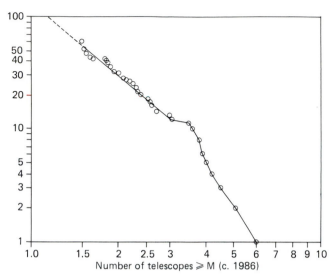

Fig. 8.6. The number of telescopes with objectives greater than or equal to *M* meters, *c.* 1986. We include the William Herschel Telescope and the Hubble Space Telescope, but do not include the Mt Wilson 60 inch and 100 inch telescopes. Schmidt telescopes are included according to the size of their mirrors, not according to the size of their corrector plates.

favor of a plethora of telescopes of different sizes at the same site. (Kitt Peak has 18 telescopes.) The telescopes listed in Table 8.1 are hardly alone on their respective mountaintops. This makes good economic sense, for it costs just as much to build a road to the top of a mountain for one telescope as for many.

Figure 8.6 gives the number of optical and infrared telescopes greater than or equal to a given effective size, as of 1986, including such anticipated telescopes as the William Herschel Telescope on La Palma and the Space Telescope, but excluding the Mt Wilson 60 inch and 100 inch, which are in the process of being closed down. There are presently some 63 telescopes with effective objective diameters greater than or equal to 60 inches (1.5 m). It can be estimated that there are about 70 telescopes in the 1.0–1.5 m range, and more than 200 professional telescopes smaller than 1.0 m since every one of the 378 observatories mentioned in the *Astronomical Almanac* must have at least one telescope. Finally, professional level astronomical observations can be carried out with telescopes as small as a 6 inch, suitably equipped with a photoelectric photometer, and many amateurs are now making a significant collective contribution with their many instruments. With all these ground-based telescopes, plus airplane- and balloon-borne telescopes and satellites, we can expect that the number of telescopes having

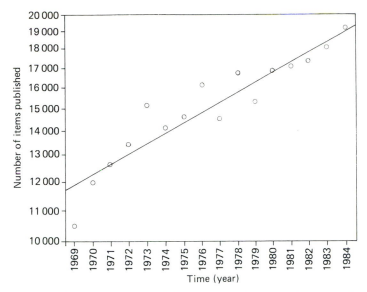

Fig. 8.7. World totals of astronomical output as a function of time, as given in the volumes of *Astronomy and Astrophysics Abstracts*.

been built continues to grow exponentially.

We can also ask: 'How many active astronomers are there?'. This can be estimated from the publication totals in the whole field. In the last 5 year cumulative volume of *Astronomy and Astrophysics Abstracts*, covering the years 1979 to 1983, a total of 83 724 published items are referenced, having been authored by 35 081 authors, a tiny fraction of which are actually the same authors counted more than once because of variations in the forms of the names on the various title pages (e.g. D. Allen and D. A. Allen). The number of published items includes a number of non-research items such as book reviews, obituary notices and errata; the same paper may be counted as a preprint and again later as the official published paper; but the reader can get a feel for the general numbers. The number of items published has been increasing exponentially with a doubling time of 22 years (see Fig. 8.7). Assuming that non-research items and duplicate papers each constitute approximately the same fraction of the total as time goes on, that means that real research items also double in number every 22 years.

Not only do the principal journals tend to increase in page size and number of shelf feet per year, reflecting the exponential growth of total research, but the number of journals published which contain astronomical research results has an exponential growth, with a doubling time of about 15 years.[14] In astronomy there has been a decrease of such numbers as the *Publications of the X Observatory*, but the appearance of new periodicals has more than

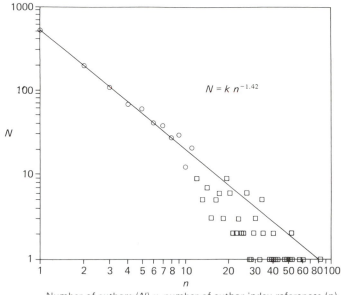

Number of authors (*N*) *v.* number of author index references (*n*)

Fig. 8.8. **The number of astronomers (N) with surnames beginning with 'A' publishing *n* items during the years 1979–83. The straight line is a fit to the first 11 points. This relationship is known as 'Lotka's Law of Scientific Productivity' and shows that the production rates of scientists are governed by a power law distribution. The principal parameter is the slope of the line plotted in a log–log graph. The steeper the slope, the greater percentage of authors publish only a few items over any time frame (a few years or lifetime totals).**

compensated. Whereas in 1969 the number of research astronomy serials and observatory contributions was about 400, by now it is on the order of 1000. The core astronomy journals (giving us about half of the total number of research papers) would amount to about $\sqrt{1000} \approx 32$ in number, of which any astronomical institution's library would have most or all.[15] It is an easy matter to determine which journals these are (*Astrophysical Journal, Monthly Notices of the Royal Astronomical Society, Astronomy and Astrophysics,* etc.)

Collaboration in scientific research is becoming more and more the rule. Scientific experiments become more complicated as time goes on, and there is more pressure for team research. The result is that the average number of authors per paper continues to rise.[16] While single-authored papers were the rule at the turn of the century, the average astronomical research paper now has two authors. Some have 30 or more.

If we investigate the productivity patterns of astronomers, we find some interesting results. One might assume that each astronomer is in the business to stay and will produce results at his or her own rate. This would give some prolific authors and some not so prolific authors, with some sort of bell-shaped curve describing productivity. In actuality this is completely wrong! If we consider the number of astronomers who write only one paper over some time interval, that will end up being the most common amount of productivity. The number of astronomers who write two papers will be substantially smaller, the number who publish three papers still smaller, and so on (see Fig. 8.8). The mathematical distribution is a power law of the form

$$N = kn^{-\alpha},$$

where N is the number of authors publishing n papers over some time interval, and α is some constant > 1. This mathematical form has the property that it does not matter if we consider a short time interval (say, 5 years) or the lifetime totals of a group of authors. The statistical phenomenon is not limited to astronomy, but holds as well for physicists, chemists, mathematicians, and other groups of researchers (each with somewhat different values of α).[17] A power law results if we bin the citations by the author index listings (where a paper co-authored by several people is counted as a paper published by each author) or attribute the papers to the senior authors only (as did the original discoverer of this statistical law). The question of fractional authorship is more complex, but I would not be surprised if a power law holds (perhaps with a slightly different value of α) if we binned the authors by 'normalized' counts (e.g. divide each paper 'event' by the number of authors and count the number of authors publishing $0+$ to 1 paper, $1+$ to 2, etc.).

There are two simple results associated with the power law distribution of scientific productivity: (*a*) the average productivity is low; and (*b*) a small fraction of the most prolific authors generates a large fraction of the total number of scientific papers.

The average productivity is obtainable from straightforward arithmetic:

$$A = \text{total number author index listings}/\text{total number authors}.$$

For the 1979 to 1983 cumulative volume of *Astronomy and Astrophysics Abstracts* this is 4.85 items per author. This is a weak function of α. For groups whose productivity is described by a power law with index $\alpha \approx 2.0$, the average number of published items per author is about 3.5.[18] A is more strongly affected by the increasing amount of multiple authorship, but the important thing to note here is that the average productivity is low even though there are many astronomers who are very prolific.

We find that just under 11% of astronomers generate more than 50% of all author index listings. Thus, the 'core' group of astronomers (both theoreticians and observationalists) over the years 1979 to 1983 amounts to 3700 people out of the total of 35 000. We find that about 7300 of our 35 000 authors generate two-thirds of the author index listings,* leaving a much larger number of non-prolific astronomers who are either under no pressure to publish, are just getting started in the business, or who will leave the field after their modest (but cumulatively significant) contributions.

It is necessary to say a few words here about actual results astronomers are obtaining. For it will hardly do to say that a paper such as 'Further photometry of the star XY Obscuratum' is equal in value to Maarten Schmidt's demonstration of the large red shift of 3C 273 (demonstrating the implied large distances, hence luminosities and ages, of quasars) or Penzias and Wilson's proof of the 3 K background radiation (evidence for the Big Bang). What are some of the most significant discoveries of the second half of the twentieth century? The two just-mentioned certainly are considered so, as well as other results presented in Lang and Gingerich's *Source Book in Astronomy and Astrophysics, 1900–1975*, along with the discovery of other ring systems in the outer solar system, gravitational lenses, and evidence of the object(s) in the very center of our galaxy, but we cannot know which of these results will retain such high ratings in the long run. A couple of examples from the past will suffice to illustrate my point.

At the end of the nineteenth century the German astronomer Friedrich Küstner and the American S. C. Chandler demonstrated that the poles of the Earth are not fixed, that each place on Earth experiences variations of latitude amounting to a couple tenths of a second of arc (several linear meters) with time scales of 12 months (the annual component) and 14.2 months (the Chandler component).[19] In 1897 this finding was described by Simon Newcomb, the foremost positional astronomer of his day, as 'the most remarkable feature of the astronomy of the last twenty years of our century'.[20] In retrospect the advent of photography and the promise of results from spectroscopic investigations have been more significant than such aspects of positional astronomy.

---

* This can be compared to the International Astronomical Union (IAU) membership of 6000, that of the American Astronomical Society (4100), and that of the Russian All-Union Astronomical-Geodetical Society (VAGO, 8000).

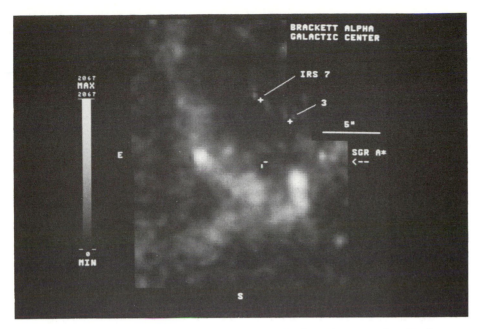

Fig. 8.9. The center of our galaxy at a wavelength of 4.05 microns, showing the distribution of atomic hydrogen gas. This image was obtained at the NASA Infrared Telescope Facility at Mauna Kea using the University of Rochester 32 × 32 infrared array. Photograph courtesy of W. Forrest and J. Pipher.

Not all results stemming from astrophysical methods which were highly regarded are still well known. I am reminded of the Tikhov–Nordmann effect,[21] which is an actual observational fact relating to the eclipse times of binary stars and how the time depends on the color of one's photographic or photoelectric filter. It was attributed to different values of the speed of light for different colors. This effect is now known to be intrinsic to the stellar atmospheres involved rather than due to any interstellar dispersion effect. G. A. Tikhov and C. Nordmann were both awarded prizes for their joint discovery, yet not many people today have heard of it.

Even the terminology in modern astronomy is rapidly changing. Quasi-stellar radio sources are now referred to consistently as quasars. Rapidly pulsating radio sources, or pulsars, were once referred to tongue in cheek as LGMs for Little Green Men, because the extreme regularity of the pulses was considered attributable to signals generated by intelligent life elsewhere in the universe. Will the newly accepted terminology include 'sosies', 'blazars',

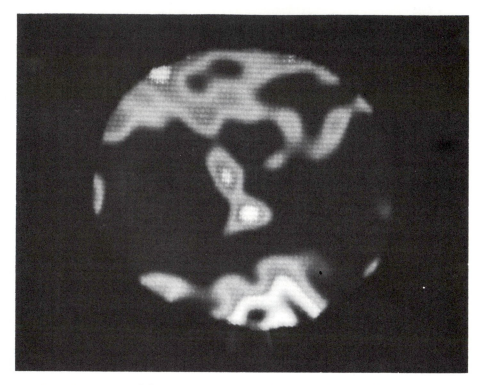

Fig. 8.10. Image of the supergiant star Betelgeuse made with the Kitt Peak 3.8 m Mayall reflector using speckle interferometry. The blotchy areas are believed to be large convective regions in the star's photosphere. Photograph from the National Optical Astronomy Observatories.

'liners', and 'noisars' – terms that have already appeared in the literature?[22]

With 20 000 astronomical papers now being published each year, it is not even possible to list the areas of investigation here, let alone the highlights. Some representative results are given on pages 211–4. Modern observational astronomy is exemplified by the push for high spatial and spectral resolution, the push to fainter limiting sensitivities, and the use of new instruments and techniques. Opening up new wavelength regimes (radio, infrared, ultraviolet, X-rays and gamma rays) has required radically different technology, the use of airborne telescopes and also satellites.

In the next sections we will discuss three premier ground-based sites and the unfolding of new windows of the universe.

Kitt Peak National Observatory

The summit of Kitt Peak, southwest of Tucson, Arizona, presently supports the largest collection of professional telescopes north of the equator. A

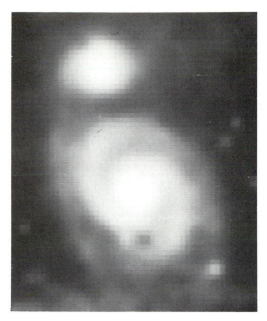

M51

Fig. 8.11. The Whirlpool Galaxy M51 in the constellation Canes Venatici, at a wavelength of 2.2 microns. The resolution of the image is 20 arc seconds. What is shown is primarily the distribution of red giant stars. The image was obtained with the United Kingdom Infrared Telescope at Mauna Kea. Photograph courtesy of Ian Gatley.

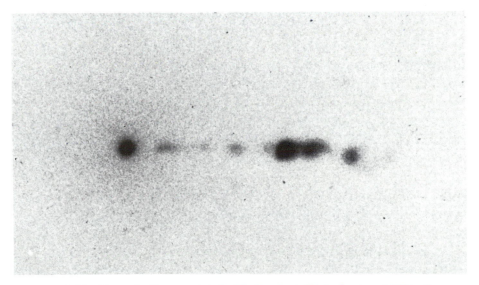

Fig. 8.12. The optical jet associated with the giant elliptical galaxy M87 in the Virgo cluster. This photograph was obtained with the Canada–France–Hawaii Telescope at Mauna Kea and exhibits a resolution of 0.6 arc seconds. Photograph courtesy of Gerard Lelièvre.

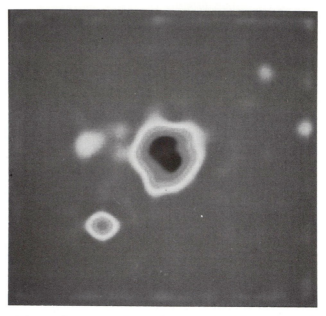

Fig. 8.13. 'The triple quasar' PG1115 + 08, which is an example of the gravitational lens effect, whereby an invisible galaxy inbetween the observer and a distant quasar bends the light of the quasar around its edges, forming multiple images of the quasar. This image was obtained with the University of Hawaii 2.2 m telescope at Mauna Kea and exhibits a resolution of 0.35 arc seconds, testifying to the excellent seeing at the 13 800 foot summit of Mauna Kea. The brightest quasar image is resolved into two components for the first time in this picture. Photograph courtesy of J. Patrick Henry and the University of Hawaii.

number of different institutions operate telescopes at this mountain: Steward Observatory of the University of Arizona (90 inch, 36 inch, and 20 inch telescopes and a 1.8 m transit telescope); Case Western Reserve University's Burrell-Schmidt (24/36 inch); McGraw-Hill Observatory (94 inch, 50 inch) of the University of Michigan, Massachusetts Institute of Technology, and Dartmouth; the National Radio Astronomy Observatory's 36 foot radio telescope; and, of course, the telescopes of the Kitt Peak National Observatory (KPNO) – two 16 inch reflectors, two 36 inch reflectors, a 50 inch reflector, an 84 inch reflector, the Mayall 158 inch reflector, and three solar telescopes, the largest of which, the McMath solar telescope, is the world's largest.[23,24]

The idea of an American national observatory had its start in a 1952 article by John B. Irwin published in *Science*. Irwin called for the establishment of a permanent observatory located in the desert and dedicated to photoelectric photometry. In 1953 a photoelectric photometry conference attended by 35 participants was held in Flagstaff, Arizona (the location of

Lowell Observatory). Among them was Leo Goldberg of the University of Michigan, who, speaking for Walter Baade of Palamar Observatory pointed out that much of what we know of extragalactic astronomy was derived from observations carried out during California's bad observing season; there was clearly a need for a major observatory whose weather was out of phase with that of California's. Goldberg proposed that the conferees not limit their thinking to photometry, but think instead of a national observatory where astronomers from any institution could come to carry out spectroscopic, photographic, and solar work as well. Albert Whitford of the University of Wisconsin also explained that the existence of such a national facility would be good for producing a better national crop of astronomers, for graduate students would not be discouraged from entering the field owing to a lack of institutions having their own telescopes (such as Harvard, Caltech, or the University of Chicago); it would be possible to have a teaching job at a smaller university and still have ample opportunity to carry out first-class research at a good site with first-class equipment.

An advisory panel for a National Astronomical Observatory was formed in 1954, consisting of Ira Bowen, Bengt Strömgren, Otto Struve, Albert Whitford, and chaired by Robert R. McMath. McMath was a civil-engineer-turned-astronomer who founded the McMath–Hulbert Observatory in Michigan, which was primarily dedicated to solar studies. He was known as a good organizer of scientists and technicians and a natural born manager. After deciding that the time was ripe for a national observatory the panel recommended that site surveys be undertaken right away. Primarily under the direction of Aden Meinel and Helmut Abt, a total of 150 mountaintops were selected as possible sites. These mountains were located in California, Arizona, New Mexico, Nevada, Utah and Texas. The desired site would have a high percentage of clear nights, experience low precipitation, not be so high that severe mountain conditions were encountered, and the site would be near a supporting city with a major university. After 2 years of further testing, Kitt Peak and Hualapai Mountain (near Kingman, Arizona) were selected for final tests.

In the years following the Second World War it was realized that to carry out science meant a shift from little science to big science. No longer could we be committed to scientific progress and rely on home-brew experiments by isolated experimenters, or expect more grandiose programs funded by rich, private benefactors. The federal government would have to become involved. McMath's committee estimated that $12 million would be needed to build the new observatory.

No single event was more significant for the realization of plans for all types of American big science than the launching of the Soviet Union's first satellite in October of 1957. Like a triple star whose components swung around a common center, the concepts of scientific progress, the hoped for technological offshoots and American national pride led the United States government to allocate unprecedented funds for the realization of scientific and technical endeavors.

The wheels had already been turning for the establishment of a national observatory. A year before Sputnik's launch, the National Science Foundation had agreed to the notion of a national observatory and asked for a proposal from a group of universities 'organized for the purpose of managing and operating an astronomical observatory facility'. In late October of 1957 the Association of Universities for Research in Astronomy (AURA) was born, consisting of seven universities: California, Chicago, Harvard, Indiana, Michigan, Ohio State and Wisconsin. Since then ten other American universities have become part of AURA: Arizona, Caltech, Johns Hopkins, MIT, Princeton, the University of Texas at Austin, Yale and the Universities of Colorado, Hawaii and Illinois. In addition to having overseen the construction of Kitt Peak National Observatory (as it came to be called), AURA also manages Cerro Tololo Inter-American Observatory in Chile, Sacramento Peak Observatory (near Sunspot, New Mexico), and the Space Telescope Science Institute (at Baltimore, Maryland). It must be emphasized that the users of the national facilities do not have to belong to AURA universities. Many foreign astronomers have also observed at the three observatories.

By March of 1958 AURA had selected Kitt Peak as the desired site for the national observatory. Permission to use the site had to be obtained from the Papago Indians, on whose reservation the mountain was located. (The mountain was a favorite dwelling place of one of the Papago gods, I'itoi.) Though it took several months for astronomers to articulate their desires to the Papagos, the actual negotiations with them were not lengthy.[25] On 28 October 1955 Papago representatives had visited Steward Observatory on the University of Arizona campus and looked at the Moon and planets through the 36-inch telescope there. By December of that year the initial access for astronomical site testing of Kitt Peak was approved by the Papago Schuk-Toak District Council and the Tribal Council. Following the selection of Kitt Peak, the Schuk–Toak District Council approved negotiation of a lease on 5 March 1958, which the Papago Tribal Council approved on 7 March. The lease for the mountain was signed by the Papago Tribal Chairman on 3 October 1958 and by the National Science Foundation on 24 October. It holds 'as long as the land is used for astronomical study and research and related scientific purposes'.

Two 16 inch telescopes had been used at Kitt Peak and Hualapai Mountain for site survey purposes. They both came to be located at Kitt Peak following the end of the initial site surveys. The first dome to be erected at Kitt Peak housed one of these telescopes, but in 1961 one of these instruments was transferred to Cerro Tololo for site testing there, the other 16 inch following a couple years later. Today there are still two 16 inch telescopes at Kitt Peak, known as the Number 3 16 inch and number 4 16 inch, but as of 1986 they are in the process of being phased out.

The first major telescope at Kitt Peak was a 36 inch reflector, which was installed in 1960. A second went into operation in 1966. In the meantime the 84 inch telescope came on line in 1964. It has a Pyrex mirror cast by Corning Glass Works. The f/2.6 primary is ground to a Ritchey–Chrétien figure, which is a fourth order curve (a parabola is a second order curve) used to give coma-free images over the whole field of view. While not primarily intended for photography at the Cassegrain focus, the primary mirror's figure allows for wider angle photography of good quality. The 84 inch telescope was primarily intended to be used for photometry and spectroscopy at the f/7.6 Cassegrain focus. Unlike the smaller telescopes at Kitt Peak, the 84 inch operates at coudé focus as well. The original cost of the 84 inch telescope alone (not including the dome and auxiliary instrumentation) was just under $1 million, which was borne by the National Science Foundation.[26]

The McMath committee of 1954 had recommended that $3.5 million be spent on a very large solar telescope. The National Science Foundation agreed to the resulting proposal under the condition that the telescope would be situated at the same site as the national observatory. This could have become a serious matter, for, as we have already mentioned, Hale had found that the daytime seeing is not necessarily good at a site where the night seeing is good. But Kitt Peak proved to be like Mt Wilson in that it enjoyed excellent daytime and nighttime observing conditions.

The McMath solar telescope (Fig. 8.14), as it came to be known, became operational 10 months after the death of Robert McMath in January of 1962. Originally, it had an 80 inch heliostat mirror; presently it has an 82 inch mirror. (A heliostat system employs one flat mirror which sends light toward the north or south celestial pole. The light path is at an angle to the horizontal equal to the latitude of the observatory. In a coelostat system two flat mirrors are used to make the light path horizontal, like the Snow telescope, or vertical, like the Mt Wilson 60 foot and 150 foot solar towers. This extra reflection in the coelostat system leads to some light loss, and the coelostat gives a 'noon shadow' of one mirror on the other. In the heliostat

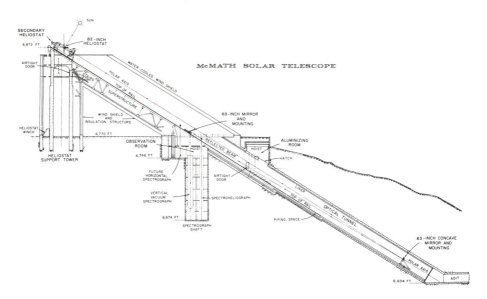

Fig. 8.14. The McMath solar telescope at Kitt Peak. Diagram from the National Optical Astronomy Observatories.

the image rotates once in 24 hours, but the elliptical cross section of the beam is constant for a day's observing. With a coelostat the image does not rotate.)

The light reflecting off the heliostat mirror 100 feet off the ground then passes almost 500 feet down the shaft of the McMath telescope. The light bounces off an off-axis parabolic mirror and passes back up the shaft to another flat mirror, from which it passes into one of a variety of spectrographs. The original second and third mirrors were made from surplus quartz blanks cast in 1931 by General Electric for their tests which were intended to lead to a 200 inch quartz blank for the Palomar telescope. (That did not come to pass, as we have related.) When the quartz blanks were abandoned they were picked up by McMath at the University of Michigan. The original concave mirror, which forms the image of the Sun, was 60 inches in diameter; the present one is 63 inches in diameter. The original second flat was 48 inches in diameter; the present one is 63 inches in diameter. The old General Electric blanks were never intended to be permanently used. They were replaced by 1966.

A principal problem with the solar telescope was the problem of air currents due to heating of the building by the Sun. To solve this the 34 000 square feet of the outer skin of the telescope is cooled by circulating 18 000 gallons of antifreeze through copper pipes. The skin of the telescope is also copper and is painted with titanium oxide paint, which is a good reflector of

Fig. 8.15. Aerial view of Kitt Peak. Photograph from the National Optical Astronomy Observatories.

light. All this conducting metal on a mountaintop naturally makes the McMath solar telescope an excellent lightning rod.

What makes the McMath solar telescope so remarkable is its extremely long focal length. This gives a solar image larger than that in any other solar telescope in the world. The Sun's disk ranges from 33.02 to 34.08 inches in diameter at the focus of the telescope (it depends on the Earth's distance from the Sun). Microphotometry of individual granules in the solar photosphere can be carried out. These granules are about 1 arc second in size, or about 0.44 millimeters at the telescope's focus. A solar spectrum 70 feet long can be obtained with the McMath telescope, allowing the Sun's spectrum to be studied with an unrivalled degree of detail.

Not only was the McMath telescope first used to confirm the existence of three of the 70 observed elements in the Sun (holmium, flourine, and chlorine), but it can be used for studying the brighter nighttime objects such as planets and bright stars. It is a dream telescope to the solar astronomer, an engineering marvel that also gives Kitt Peak its unique profile (Fig. 8.15).[27]

The most powerful and by far the most expensive ($10 million) telescope at Kitt Peak is the 158 inch Mayall reflector (Fig. 8.16), whose twin resides at Cerro Tololo Inter-American Observatory (Fig. 8.3). The Mayall reflector has

Fig. 8.16. Dome of the 3.8 m Mayall reflector, Kitt Peak. Photograph copyright
by Michael Broyles, used by permission.

an f/2.8 Ritchey–Chrétien primary made of fused quartz. General Electric used its previous experience in quartz mirror making to cast the 15 ton disk. It is the only telescope at Kitt Peak large enough to have a prime focus cage. The Cassegrain optical system is f/8, while at coudé the light cone is f/190, thus giving an enormous effective focal length greater than 700 m, far longer than the tower telescopes erected on the Marly tower in Paris in the seventeenth century (see Fig. 4.5).

In the 1960s, American astronomers rated as one of the highest priorities the construction of a number of large telescopes. While visiting astronomers could obtain telescope time at Palomar and Mt Wilson, those telescopes were becoming increasingly demanded by their own staff members. With the construction of the Kitt Peak and Cerro Tololo observatories there were at last more big telescopes to go around. However, demand often outsteps supply, and the large telescopes of the world (not just at Kitt Peak and Cerro Tololo) are often oversubscribed by a factor or three or more. A large majority of observing proposals are rejected.

Has Kitt Peak been a success? The answer is, not surprisingly, yes. While the public may be more interested in the world's largest nighttime telescope (as Palomar became after many years of planning and construction), Kitt Peak cannot make such a boast. However, Kitt Peak does have the world's longest focal length solar telescope, thus giving the greatest resolution. The Mayall telescope has been used for many projects, among them speckle studies of Betelgeuse (Fig. 8.10), the discovery of giant luminous arcs between galaxies,[28] the discovery of the high velocity wind in the very center of our galaxy,[29] which is evidence for the existence of a massive black hole or other unusual source(s), and the telescope has allowed the exploitation of the excellent seeing at Kitt Peak for high spatial resolution deep sky images with CCD (charge-coupled device) detectors. The smaller telescopes are periodically updated with new instrumentation and computer facilities as well. They too have been enlisted to contribute to such 'big telescope projects' as the distribution of galaxies and the expansion rate of the universe,[30] and the discovery of the first evidence of the gravitational lens effect.[31]

The excellent seeing at Kitt Peak and the effective support facilities there have led to the establishment of observatories by other institutions, namely the McGraw-Hill Observatory of Michigan–MIT–Dartmouth, Case Western Reserve's Burrell-Schmidt, Steward Observatory, and the National Radio Astronomy Observatory (NRAO) radio telescope. Thus, optical, infrared, and radio astronomy are all carried out at this busiest of astronomical mountaintops.

Mauna Kea

The summit of Mauna Kea on the Island of Hawaii is regarded by many to be the premier astronomical site in the world. While in bad years the fraction of clear nights might be under 50%, in good years it is as cloudless a site as can be found. More important than this is its altitude. At 13 800 feet (4205 m) Mauna Kea is almost twice as high as any other major observatory site in the world.* Though observers there do not have to deal with lightning storms, rattlesnakes, scorpions, or hordes of moths (as at Kitt Peak), the weather at Mauna Kea can be extremely severe. One can experience rain, sleet, snow, hail, freezing fog and sunshine all on the same *summer's* day. Winter storms can batter the summit with winds in excess of 120 miles per hour. One of the first site testers at Mauna Kea thought he was going to meet the Abominable Snowman when he heard tremendous scraping noises on the outside of his trailer, only to find that the wind was peeling the paint right off the side of his shelter! Yet the altitude of Mauna Kea can be a blessing as well. The extreme altitude places the telescopes above much more of the Earth's atmosphere than at other sites, allowing observations at infrared and submillimeter wavelengths that cannot be carried out elsewhere.

While astronomical activity has been going on at Mauna Kea only since 1964, the mountain has had visitors for centuries. The ancient Hawaiians regularly went to the adze quarry (Keanaka'ko'i) at the 12 000 foot level in order to fashion stone tools. Occasionally, the remains of deceased relatives were buried at the mountain, and there are reports of certain families depositing the umbilical cord of newborn babies in Lake Waiau at the 13 000 foot level.[32] The summit of Mauna Kea ('white mountain') is the home of Poli'ahu, the snow goddess. The fire goddess, Pele, resides at Mauna Loa ('long mountain') 30 miles to the south.

The first recorded visit to the summit of Mauna Kea was by the Rev. Joseph Goodrich[33] on 25–26 August 1823.[34] Goodrich travelled on foot from Waimea (elevation 2500 feet) up the northwest (steepest) face of the mountain. (Most early ascents of the mountain were subsequently made by travelling west from Hilo across the saddle between Mauna Kea and Mauna Loa, then north to Mauna Kea's summit.) Except for the early part of his hike, Goodrich went to the summit alone, supplied only with a coat, blanket, some food, and fire making implements, even though he believed that the summit was at an elevation of 18 000 feet. Moonlight showed him the way to the very summit. He wrote in April of 1825:

* The only place comparable is Pic du Midi in the French Pyrenees (2865 m/9400 feet). There the biggest telescope is a 2 meter. See end note 87 to this chapter.

The first time I was at the highest peak about three o'clock at night, in the month of August [1823]; the thermometer stood at 27 deg, 5 below the freezing point. I passed over several banks of snow, that lay to the northward of the highest peaks ... and the change was so great in passing from a torrid to a frigid zone, that I was under the necessity of travelling all the time I was up there to prevent freezing. The second time that I ascended was in April last [1824]. There appear to be three or four different regions in passing from the sea shore to the summit. The first occupies five or six miles, where cultivation is carried on, in a degree, and might be to almost any extent; but as yet, not one twentieth part is cultivated. The next is a sandy region, that is impassable, except in a few foot paths. Brakes, a species of fern, here grow to the size of trees; the bodies of some of them are eighteen inches in diameter. The woody region extends between ten and twenty miles in width. The region higher up produces grass, principally of the bent kind. Strawberries, raspberries, as large as butternuts, and whortleberries flourish in this region, and herds of wild cattle are seen grazing. It is entirely broken up by hills and valleys, composed of lava, with a very shallow soil. The upper region is composed of lava of almost every form, from huge rocks to volcanic sand of the coarser kind. Some of the peaks are composed of coarse sand, and others of loose stones and pebbles. I found a few specimens that I should not hesitate to pronounce fragments of granite. I also found fragments of lava, bearing a near resemblance to a geode, filled with green crystals, which I suppose to be augite.

Very near to the summit, upon one of the peaks I found eight or ten dead sheep; they probably fled up there to seek a refuge from the wild dogs; I have heard that there are many wild dogs, sheep and goats. Dogs and goats I have never seen.[35]

At the summit Goodrich had found a stone cairn erected by some former visitor. There were also wild cattle to be found on the middle slopes of the mountain, having been brought by the Europeans in the late eighteenth century and which multiplied greatly because the Hawaiian chiefs had at first forbidden the natives to hunt them. (Today wild sheep and wild boar are still to be found at the 6000–10 000 foot level.)

Goodrich made other ascents, accompanying the botanist James Macrae of *HMS Blonde* to the summit on 16–17 June 1825;[36] regarding an ascent of December 1831 Goodrich relates: '...nothing occurred very materially different from what I have heretofore mentioned; a severe headache, affecting the natives as well as myself, with sickness at the stomach and vomiting of bilious matter, usually attends me in those lofty regions'.[37]

The botanist David Douglas ascended Mauna Kea in January 1834, finding it to be 13 851 feet tall, much lower than previous estimates. He relates:

> While on the summit I experienced violent headache, and my eyes became bloodshot, accompanied with stiffness in their lids.
>
> Were the traveler permitted to express the emotions he feels when placed on such an astonishing part of the earth's surface, cold indeed must his heart be to the great operations of Nature ... Man feels himself as nothing – as if standing on the verge of another world.[38]

Douglas did not make multiple visits, as he died in July of 1834 after falling into a pit designed for trapping wild cattle.

Two other early expeditions were the one led by James Jarves in July of 1840,[39] and that of the US Exploring Expedition in January of 1841.[40] Almost a century later another expedition leader wrote: 'As a result of these various reconnaissance trips, the part of Hawaii above the 10,000-foot level became almost as well known by 1841 as it was down to ten or fifteen years ago.'[41]

It should be clear from these early accounts that a trip to Mauna Kea is a trial for body and soul. This was true when the mode of transport was by foot, horse, or mule, and is still true today with four-wheel drive vehicles. Altitude sickness is a common problem, though it has been found that most people become acclimatized if they do not descend below 8000 feet during their stay. By the third or fourth day, spending no more than 12 hours at the summit, one feels quite functional, though one's thinking ability is *always* worse at the summit compared to one's sea level performance.[42] That is why the astronomers who work at Mauna Kea stay at a mid-level facility at Hale Pohaku ('house of stone') at the 9200 foot level.

A case can be made that the establishment of a permanent mountaintop observatory at Mauna Kea came about as a result of a powerful tidal wave that devastated Hilo on 23 May 1960. It severely impacted the perennially shaky Big Island economy. In the aftermath of this natural disaster various 'hair-brained' schemes were hatched to rejuvenate the local economy, including the marketing of lava and the possibility of developing astronomical facilities.[43] At the suggestion of Howard Ellis of the US Weather Bureau's Mauna Loa Observatory, in 1963 the Executive Secretary of the Hawaii Island Chamber of Commerce, Mitsuo Akiyama, blanketed US and Japanese universities and research organizations with information on the possibilities of developing Mauna Kea and Mauna Loa as astronomical sites.[44]

Already in 1956 the University of Hawaii had established a solar observatory at Haleakala on Maui (the 'house of the Sun', elevation 10 000 feet). In that same year the US Weather Bureau had established at Mauna Loa an atmospheric research station (elevation 11 134 feet). Most of Akiyama's letters went unanswered, but he did receive an enthusiastic reply from Dr Gerard Kuiper of the Lunar and Planetary Laboratory of the University of Arizona. (Kuiper, the former director of the Yerkes and McDonald Observatories, and discoverer of the fifth moon of Uranus and the second moon of Neptune, moved to Arizona in 1960 to set up the Lunar and Planetary Laboratory.) Kuiper had already set the wheels of motion in action for just such a program as Akiyama envisioned. From October 1962 to August 1963 NASA and the Advanced Research Program Agency of the Department of Defense had funded a site testing program on Haleakala, the result of which was the establishment of three more telescopes there at a cost to the Universities of Hawaii and Michigan of $4.3 million.[45] But Kuiper wrote:

> Our tests on Haleakala made it abundantly clear, however, that a distinctly higher site, in the same general area, than Haleakala would have had very marked advantages. Haleakala is not high enough above the average top of the cloud layer surrounding the Hawaiian Islands with the result that night observations are frequently interrupted; though it is true that during a fair fraction of the nights the conditions on Haleakala are very good.[46]

Kuiper had flown over Mauna Kea in June of 1963 and he did so again in January of 1964 as part of a serious effort to put together a plan for site testing on Mauna Kea, which was favored over Mauna Loa because of the lower expected occurrence of seismic and volcanic activity.[47]

One of the biggest problems of any Mauna Kea site was its accessibility, so one of the first orders of business was to get some sort of road to the summit area. On 17 January 1964 Kuiper met with Hawaii Governor John Burns in Honolulu and was assured that there would be no problem in getting $25 000 in State funds for the road. (NASA could not fund the road because their funds were designated for research.) That same day Kuiper met with University of Hawaii President Thomas Hamilton and Dr George Woollard, the Director of the University's Institute of Geophysics, to solidify a commitment for cooperation between the Universities of Hawaii and Arizona, for which they had a precedent with the site testing on Haleakala. That evening Kuiper drew up a proposal which called for the establishment of a $12\frac{1}{2}$ foot dome and a test telescope (later determined to be a $12\frac{1}{2}$ inch reflector) at the summit of the third highest cinder cone,[48] the so-called Goodrich Cone (elevation 13 440 feet).[49] Lyman Nichols, a State District Wildlife Biologist, hiked to the summit in February of 1964 to survey the area. On the basis of Nichols' photographs and comments, Kuiper decided instead that Pu'u Poli'ahu ('cinder cone of the snow goddess', elevation 13 631 feet) would be a better site.[50]

On 26 February 1964 the Board of Directors of the Hawaii Island Chamber of Commerce approved a recommendation of its Legislative Committee to support the ideas of Kuiper and President Hamilton that a portion of Mauna Kea be set aside as a scientific preserve. This proposal was brought up at a session of the State Legislature the following month.[51] The idea was approved by the Legislature on March 25th and sent to the Governor to begin further work on the formal establishment of a Mauna Kea Science Reserve.[52]

Meanwhile, Fujio Matsuda, the Director of the State Department of Transportation, told Governor Burns that the road would cost $42 000 instead of the originally estimated $25 000.[53] This difference was not considered a serious matter.[54] The road was to head up out of Hale Pohaku at 9200 feet, between Lake Waiau and Goodrich Cone at 13 000 feet, and up to Pu'u Poli'ahu, a distance of 5.9 miles. (The present road from Hale Pohaku to the summit – which is not much beyond Pu'u Poli'ahu – is 8.3 miles long. The original road did not include the switchbacks below 11 000 feet, and hence was very steep in places.) On 13 April 1964 a bulldozer headed out of Hale Pohaku to begin the road.[55] It was finished a month later.[56]

In the span of 11 days (starting 1 June) the concrete foundation for the dome was laid using water from Lake Waiau,[57] the $12\frac{1}{2}$ foot Ash Dome was installed, and a $12\frac{1}{2}$ inch telescope was set up.[58] Instead of the mirror that came with the telescope, made in the optical shop of the Lunar and Planetary Laboratory, Kuiper's intended observer, Alika Herring, installed his own mirror, which was regarded by opticians who tested it to be perhaps the finest $12\frac{1}{2}$ inch mirror ever made.[59] Herring was an optician and amateur astronomer who had impressed Kuiper with the quality of his lunar and planetary drawings. He was involved in site testing for a number of years during the 1960s, including the testing on Maui in 1962–3 and tests in the American southwest and Chile.[60]

Now, on Mauna Kea, which it was hoped to be a superb site, an experienced observer with a quality telescope was ready to test the sky. Observations began on Saturday, 13 June 1964.[61] The observations consisted of lunar and planetary drawings, tests of the atmospheric seeing and transparency, and measurements of the atmospheric water vapor content.

The plan was to get the dome on the mountain, get the telescope working, and take some observations before the formal dedication. This took place on

Fig. 8.17. Dedication of the Mauna Kea Observatory, 20 July 1964. The scene is at Hale Pohaku, at the 9200 foot level, where the present mid-level facilities are. Photograph courtesy of Ewen Whitaker and the Lunar and Planetary Laboratory, University of Arizona.

20 July, consisting of a ceremony at Hale Pohaku attended by 200 people and some speeches given by Kuiper, Governor Burns, and others.[62] By this time Alika Herring had already told Kuiper, 'This mountain is it!' In Kuiper's speech on that dedication day he proclaimed, 'This mountaintop ... is probably the best site in the world – I repeat – in the world, from which to study the Moon, the Planets, the Stars ... It is a jewel! This is the place where the most advanced and powerful observations from this Earth can be made'.[63]

Just how good was the site? It was later verified

> that Herring's seeing scale (0 very poor, 3 fair, 5 good, 7 excellent, 9–10 entirely perfect for many minutes without the slightest interruption) was stable and in agreement with the customary scale used by experienced double-star and planetary observers ... [Herring's telescope] showed for a bright star the 0".36 stellar diffraction disk surrounded by at least 6 diffraction rings. These rings are extremely sensitive to seeing and give a sharp measure of it.[64]

Herring had found many nights of seeing 9–10. His lunar and planetary drawings showed exquisite detail.

Fig. 8.18. Observer Alika Herring and the $12\frac{1}{2}$ inch telescope atop Pu'u Poli'ahu, the second highest cinder cone at Mauna Kea. This was the first telescope established at the mountain, in the summer of 1964. Photograph courtesy of Ewen Whitaker and the Lunar and Planetary Laboratory, University of Arizona.

The sky at Mauna Kea is also extremely dark. Visually it is possible to detect stars of apparent magnitude 14.5 using a 6-inch telescope. Given modern electronic detectors on one of the major telescopes at the summit, one can push beyond 26th magnitude.

A more objective evaluation of the seeing conditions must be made by careful photometric determinations of the atmospheric extinction (the rate at which stars dim as they get lower in the sky) as a function of wavelength, photometry of the night sky brightness, and photographic records of the size of stellar images. An ideal site is above the local atmospheric inversion layer, giving rise to small variations of temperature at night.[65] A comparison of four major observatory sites is given in Table 8.2. Examples of the type of high-resolution imaging capable at Mauna Kea are shown in Figs. 8.12 and 8.13.

The site testing continued during the summer of 1964. Alika Herring

Fig. 8.19. M. Akiyama (in rear), and (from left to right) G. Kuiper, G. Fielder, J. Texereaux and H. Ellis at the site of the first telescope sited at Mauna Kea, 12 October 1964. Photograph courtesy of Alika Herring.

made four trips to Hawaii to conduct observing runs lasting from 2 to 6 weeks. William Hartmann, then a graduate student at the University of Arizona, also participated in the first observing run.[66] By October the first phase of the site testing was over.[67] Kuiper proposed a cooperative venture with the University of Hawaii's Institute of Geophysics, calling for the establishment of a 28-inch photometric telescope in a 20 foot dome, to be moved to Mauna Kea from the University of Arizona's Catalina Station No. 1 of the Lunar and Planetary Laboratory.[68] As opposed to the qualitative evaluation of the seeing conditions resulting from visual observations, photometry done at the mountain would allow a quantitative evaluation.

Kuiper's optimistic feelings for the anticipated results were all on the order of the following: 'Mauna Kea, Hawaii, is regarded as the most promising site on US soil and, in fact, the most promising site now known anywhere in the world'.[69] Furthermore, he declared, 'I regard the discovery of Mauna Kea as a superior observatory site one of my main professional accomplishments'.[70] This was quite a confession indeed given that Kuiper is easily considered one of the major astronomers of the twentieth century.

However, there was an increasing feeling that Kuiper was already overextended and overcommitted as far as NASA was concerned. He was

Table 8.2. *Comparison of Major Observatory Sites*[65]

Location	Nighttime weather		Hours of cloud cover			
	Mean temperature change (°F)	Relative humidity	0%	<60%	<80%	Months observed
Mauna Kea	4.0	39	67	87	89	7
	—[a]	—	58	68	—	14
Cerro Tololo	2.7	33	51	78	83	23
La Palma	—	—	60	72	—	12
Kitt Peak	5.2	64	—	—	—	8

Location	Altitude (feet)	Percentage of night with seeing[b]				Months observed	Nights observed
		<1"	1".1–1".5	1".6–2".0	>2".0		
Mauna Kea	13 796	21 (39)	29	19	30	12	217
	—[a]	9 (36)	43	11	37	6	76
	—	4 (21)	29	12	56	6	77
Cerro Tololo	[7100]	24 (29)	32	22	22	23	509
La Palma	7762	26 (32)	26	15	33	12	245
Kitt Peak	6772	15 (18)	30	16	39	29	253

[a]Multiple entries are from different surveys.
[b]Seeing values are at the elevation angle of the polestar. Values in parentheses indicate seeing reduced to an elevation angle of 54°.

even told by Dr Woollard, the director of the University of Hawaii's Institute of Geophysics: '[I]t is no exaggeration to say that the surest way of our not getting NASA support for developing an astronomical program on Mauna Kea would be association with you.'[71] The situation was summarized by Woollard as follows:

> [At NASA] there was some aversion to having Professor Kuiper associated with the Mauna Kea project despite the fact that he had been commissioned by NASA to carry out the preliminary test observations. This aversion was not related to Professor Kuiper's competence as an astronomer, but rather to the fact that since his present NASA supported program is so large [the Lunar and Planetary Lab, the Ranger Moon program, the Catalina Stations of the LPL, etc.], it was felt that he could not do justice to an additional program on Mauna Kea. It was suggested that we [the University of Hawaii] might consider several astronomers who it was felt might make a more satisfactory partner than Professor Kuiper.[72]

The University of Arizona dropped its plan to transport the 28 inch telescope to Mauna Kea, and proposed that a 60 inch telescope be built. It was to cost $900 000 and be ready in time for the 1967 opposition of the planet Mars.[73] The University of Hawaii decided to submit its own proposal for an 84 inch telescope.[74] This was submitted on 23 March 1965.[75] The University of Hawaii proposal made it clear that the telescope would not automatically be placed on Mauna Kea, that Haleakala was still in the running in the eyes of John Jefferies and his colleagues at the University of Hawaii, Manoa. The final decision was to be made on the basis of more site testing.[76]

The unusual thing about the University of Hawaii proposal is that at that time there was no department of astronomy. Four solar astronomers who did daytime astronomy at Haleakala were competing with an established observatory builder with many university and government connections.[77] In retrospect it may seem like Robin Hood (or, to be more exact, John Jefferies) and his Band of Merry Men, but hindsight also has shown us that it was correct that the development of Mauna Kea be run from Hawaii instead of Arizona. NASA awarded the contract to the University of Hawaii. The contract was signed on 1 July 1965. Kuiper felt that the mountain had been stolen from him, but later admitted that it would have been difficult logistically to run a full-blown Mauna Kea Observatory from Tucson.[78]

If one reads the two telescope proposals, the one submitted by the University of Hawaii was clearly better because it considered long-term plans. NASA felt that putting the right telescope on the mountain,

Fig. 8.20. The University of Hawaii 2.2 meter reflector at Mauna Kea. Photograph copyright by Michael Broyles, used by permission.

administered from what they considered the correct arena, outweighed the
desire to have a larger telescope at the mountain by the summer of 1967.

The University of Hawaii telescope turned out to be an 88 inch telescope
(Fig. 8.20) and cost more than $ 3 million.[79] When it became operational in
1970 it was the world's eighth largest telescope.[80]

In a book on astronomical centers of the world, Gerard Kuiper must be
regarded as one of the kings. He did not let his setback regarding Mauna Kea
slow him down. Until his death in 1973 he remained as active as ever in
his attempts to further the development of astronomical facilities.[81] He
pioneered the selection of a number of superior mountain sites which
support major astronomical facilities.[82] He is rightly considered the founder
of the Mauna Kea Observatory, and it is no accident that the first major
observatory which could *fly* higher than any mountain on Earth is named
the Kuiper Airborne Observatory (see p. 254).

On the other side of the coin, while Mauna Kea is the place that Gerard
found, it is the house that Jack built.[83] John Jefferies gathered together a

Fig. 8.21. The 3.8-meter United Kingdom Infrared Telescope at Mauna Kea.
Photograph courtesy of the Royal Observatory, Edinburgh.

group of primarily young, adventurous astronomers in the late 1960s and 1970s whose common purpose was to prove to the world that a major observatory could be built at Mauna Kea, and that they could do it. The Institute for Astronomy was established at the Manoa campus of the University of Hawaii in 1967 (Manoa is a section of Honolulu). Twenty years later it ranks as one of the major American university departments of

Fig. 8.22. The 3.6 m Canada–France–Hawaii Telescope at Mauna Kea. Photograph copyright by Michael Broyles, used by permission.

Fig. 8.23. The 3 m NASA Infrared Telescope Facility at Mauna Kea. Photograph copyright by Michael Broyles, used by permission.

Fig. 8.24. Aerial view of Mauna Kea, showing (from bottom, and proceeding along the ridge) the domes of the University of Hawaii's 24 inch telescope, the United Kingdom Infrared Telescope, the University of Hawaii's 88 inch and their second 24 inch telescope, the Canada–France–Hawaii Telescope, and (at far left) the NASA Infrared Telescope Facility. Photograph courtesy of the Royal Observatory, Edinburgh.

astronomy. Their astronomers get telescope time on every telescope present and future at the mountain. If astronomy wealth is measured by clear nights with state-of-the-art equipment on one of the world's best sites, University of Hawaii astronomers are reaping a lion's share of the assets. At the time of writing, University of Hawaii astronomers were getting 15% of the time on the United Kingdom Infrared Telescope (Fig. 8.21), 12% of the time on the Canada–France–Hawaii Telescope (Fig. 8.22), and 25% of the time on NASA's Infrared Telescope Facility (Fig. 8.23; 15% of the total for research, and 10% of the total for engineering). Given all the expenses of running a major telescope facility, a clear night is regarded to be worth $10 000 of operating costs. At present, the British and Dutch are building a 15 m diameter submillimeter and millimeter wavelength telescope, called the James Clerk Maxwell Telescope; Caltech is building a similar 10.4 m millimeter wave telescope. University of Hawaii astronomers will receive up to 10% of the time on those facilities. They will also receive 10% of the telescope time on the Keck Telescope (Fig. 9.2), a 10 m aperture optical and infrared telescope to be built by Caltech and the University of California.

Why are there so many major telescopes at Mauna Kea, with even larger ones on the way? Mauna Kea is almost twice as high as any other major observatory. Observations at far infrared wavelengths (20–35 microns) and at submillimeter wavelengths must be carried out at high altitude sites. While observations at optical wavelengths and many infrared and longer wavelengths can be made at lower elevations, the altitude of Mauna Kea gives it a distinct advantage over other mountains. At a latitude of 20 °N the telescopes there can cover a very large fraction of the whole sky. Mauna Kea has become an astronomer's Mecca, a modern Uraniborg. It has already amply proven itself to be a place where the most advanced observations on Earth can be made.

Astronomy in the Canaries

The islands of Tenerife, La Palma, Gomera and Hierro form the Spanish province of Santa Cruz de Tenerife. With the rest of the Canary Islands they are to be found at a latitude of 28 °N off the coast of Morocco. The official inauguration of the observatories in the Canaries took place on 28–29 June 1985, attended by kings and queens, dukes and princes, Nobel Prize winners, ambassadors and governmental representatives, and mere observatory directors, astronomers and other mortals. The telescopes (see Table 8.3) at the Observatorio Astronomico del Teide at Izaña on Tenerife (2369 m/7800 feet) and at the Roque de los Muchachos Observatory at the top of La Palma (same altitude) are operated by many respective Western European countries: France, Spain, West Germany, Belgium, Sweden, Denmark, Finland, Norway, the United Kingdom and The Netherlands. Only at Kitt Peak and at La Silla (ESO) is there such a large collection of observational instruments.

The westernmost island of Hierro in the Canaries was known in classical times, and until the establishment of the Greenwich Prime Meridian in 1884 longitudes were reckoned with respect to Hierro's position. Just as Edmond Halley surveyed the skies from St Helena in 1676, and Lacaille and John Herschel made their surveys of the southern sky from South Africa, so did Charles Piazzi Smyth come to Tenerife in 1856 to conduct the first observations from its summit. He stayed 113 days on Tenerife, half of which was spent at the summit 'enjoying' the rugged mountain weather. This is not

Table 8.3. *Telescopes in the Canary Islands*[84a]

Diameter (cm)	Description	Owner	Operative from
At Tenerife:			
30	Photopolarimeter for diffuse sources	Bordeaux Observatory (F)	1964
25	Heliograph	IAC (Sp)	1969
40	Vacuum solar telescope	Kiepenheuer Institute (G)	1972
155	Infrared flux collector	IAC (Sp)	1972
50	General purpose telescope	Mons University (B)	1972
6 × 20	6 channel photopolarimeter	IAC (Sp)	1984
80	General purpose telescope	IAC (Sp)	1985
45	Vacuum solar telescope	Goettingen University (G)	1985
60	Vertical solar telescope	Kiepenheuer Institute (G)	1985
90	THEMIS solar telescope	INAG (F)	[1990]
At La Palma:			
60	General purpose telescope	Royal Academy of Sciences (S)	1982
60/44	Heliostat/Cassegrain solar telescope	Royal Academy of Sciences (S)	1982
18	Carlsberg automatic transit circle	Copenhagen University (D)	1984
100	Astrometric telescope (Jacobus Kapteyn)	SERC/ZWO (UK/NL)	1984
250	General purpose telescope (Isaac Newton)	SERC/ZWO (UK/NL)	1984
420	General purpose telescope (William Herschel)	SERC/ZWO (UK/NL)	1987
250	General purpose telescope (Nordic Optical Telescope)	(N)	[1988]

[a]*Abbreviations:* B: Belgium; D: Denmark; F: France; G: West Germany; IAC: Instituto de Astrofisica de Canarias; N: Nordic Countries (Denmark, Finland, Norway, and Sweden); NL: Netherlands; S: Sweden; SERC: Science and Engineering Research Council; Sp: Spain; UK: United Kingdom; ZWO: Netherlands Organisatie voor Zuiver-Wetenschappelijk Onderzoek.

such a strange activity for an astronomer, but it was his honeymoon. Lest the reader think, however, that this is out of the ordinary for astronomers, consider that Wilhelm Struve interrupted his first honeymoon to go off for a job interview as Director of the Mannheim Observatory,[85] George Ellery Hale had as 'the goal of the entire [honeymoon] trip' a visit to Lick Observatory to admire the new 36 inch refractor,[86] and even this writer spent part of his first honeymoon at Yerkes Observatory. Urania, the muse of astronomy, is a rival queen in the astronomer's castle, or so it seems in many instances!

It was claimed by Edward S. Holden, the first Lick Observatory director, that mountain observatories could trace their development from 1741, when the French astronomer François de Plantade died while observing at the Pic du Midi in the French Pyrenees at an altitude of 9439 feet.[87] Others claim that it was Piazzi Smyth's observations at Tenerife in 1856 that showed the advantages of high altitude observing.[88] (At least he survived.) His geophysical and meteorological measurements, observations of the Moon, planets, and double stars, observations of the zodiacal light, ultraviolet radiation from the Sun and infrared radiation from the Moon were laid down in a 450 page book entitled *Tenerife, an Astronomer's Experiment: or Specialities of a Residence above the Clouds.*

Over the next hundred years, various astronomers sporadically came to the Canaries. In the 1890s O. Simony and K. Ångström made measurements of solar radiation. In 1910 Jean Mascart of the Paris Observatory photographed Halley's Comet and also observed the zodiacal light. Many astronomers from around the world came to the Canaries in 1959 to view a total solar eclipse, and from this date we can reckon the modern astronomical activity at the archipelago. Much like the early situations at Kitt Peak and Mauna Kea, big plans were made which hinged on carefully executed quantitative measurements of the seeing conditions (see Table 8.2 for comparison). One of the early concerns in the Canaries was the occurrence of high altitude sand from the Sahara. On the basis of 20 years of meteorological records it was found that only 2.2% of the days without cloud cover were affected by airborne sand from the African continent.

Similar to the role of the University of Hawaii's Institute for Astronomy for Mauna Kea, developments at Tenerife and La Palma are supervised by the Instituto de Astrofisica de Canarias (IAC) at the University of La Laguna. This nominally gives IAC astronomers 20% of the allocated time of the telescopes installed at both Canarian summits.

The largest telescope presently at the Teide Observatory at Tenerife is the 1.55 m infrared flux collector, originally operated by the British, but given over to the Spanish for financial reasons. It was the prototype for the 3.8 m United Kingdom Infrared Telescope at Mauna Kea. In addition to astrometry carried out at Tenerife and La Palma, the establishment of the international collective in the Canaries has meant the rebirth of observational astronomy based at the Royal Greenwich Observatory. The Isaac Newton Telescope (INT), originally a 98 inch situated at Herstmonceux from 1967 to 1979,

Fig. 8.25. Panorama of the Roque de los Machachos Observatory at La Palma in the Canary Islands. From left to right are the Danish Carlsberg Automatic Transit Circle (in the small metal shed), the dome of the 4.2 meter William Herschel Telescope (partially constructed), the tower of the Swedish solar telescope, the half-hidden dome of the 2.5 m Isaac Newton Telescope, and the 1.0 m Dutch Kapteyn Telescope. This photograph was taken in December 1984, and is reproduced courtesy of the Royal Greenwich Observatory.

saw first light at La Palma in February of 1984.[89] Even a new mirror (a 100 inch) was placed in the telescope at La Palma. Starved for some years for photons of fainter sources, British astronomers and others are now eagerly soaking them up at the business end of the INT.[90]

Largest of all the telescopes in the Canaries will be the William Herschel Telescope (WHT),[91] a 4.2 m reflector built by Grubb Parsons of Newcastle upon Tyne. (In fact, it is planned to be their last telescope.) Unlike the other 4 m class telescopes around the world, the WHT, like the Soviet 6 m telescope, will be mounted on an altazimuth mounting. This, of course, means that a computer must update the two coordinates continuously to guide on an object, and the rotation of field angle must also be calculated. Objects near the zenith cannot be easily observed but the altazimuth mounting allows for a much more easily balanced telescope. Since modern telescopes are all heavily dependent on computers anyway, the 'constraints' of the WHT are actually advantages; they have already been proven to be easily surmountable.

The WHT will become operational while this book is in press. Astronomers can then access general user facilities having two 3–4 m class instruments in Western Europe (the other being the 3.5 m telescope at Calar Alto, Spain), three in Hawaii, one in Arizona, two in Chile, and one in Australia (see Table 8.1). This is still not enough to satisfy the research needs of modern astronomers, since all of these telescopes are oversubscribed by a factor of three or more. (Each telescope time committee can allocate only one-third or less of the total number of requested nights by all potential observers.) Some of the future plans of astronomers are described in Chapter 9, but first we must briefly cover radio astronomy and astronomy carried out with satellites.

Radio astronomy

Until 50 years ago, astronomers confined their attention to a very small slice of the electromagnetic spectrum, the optical window, those light waves ranging from 3500 to 7000 Ångströms in wavelength. There are many kinds of electromagnetic radiation (see Appendix I), from the most energetic gamma rays to the least energetic radio waves. It was first at radio wavelengths that a non-optical window on the universe was opened. Until the 1950s there was little contact between the pioneers of radio astronomy and mainstream astronomers. The latter were used to taking optical spectra, photographs and measuring star positions. Radio astronomers required a great deal of electronics – something that has intimidated optical astronomers until quite recently and only ceasing with the acceptance of modern CCD (charged-coupled device) detectors as a standard imaging method. There is new scientific jargon and means of doing business in the world of radio astronomy – superheterodyne receivers, VLBI (very long baseline interferometry), synchrotron radiation, Fourier Transform theory. While astronomers 'at' different wavelengths speak different languages, they are all trying to describe the constituents of the universe regarding whatever those constituents do. Since the Second World War, there has been a new revolution in astronomy characterized by the opening up of all regions of the electromagnetic spectrum. A full understanding of the behavior and morphology of cosmic objects now requires accounting for their energy output at a multitude of wavelengths. This has entailed the study of a great variety of processes giving rise to electromagnetic radiation, such as the annihilation of particles (according to $E = mc^2$), to the rotation and vibration of molecules, shock waves, and the spiralling of relativistic electrons through interstellar magnetic fields.

A sufficient amount has been published about the early history of radio astronomy that I need not go into much detail here.[92,93,94] Karl Jansky (1905–50), a Bell Telephone radio engineer, first detected extraterrestrial radio waves while investigating sources of radio interference in order to improve the signal quality of trans-Atlantic telephone calls. His radio 'telescope' was an antenna 29 m long, operating at a wavelength of 14.5 m (a frequency of 20.5 MHz), which rotated horizontally, thus allowing the azimuth of the radio radiation to be determined. A reproduceable radio hiss was obtained from due south at a steadily earlier time each succeeding day, and over the course of a year the signal had lapsed by a whole day. Only a source in the celestial realm could do this. Jansky easily fixed the right ascension of the source of the radiation by noting the local sidereal time when the azimuth of the signal was due south. He found it to be at 18h 00 m $\pm$ 30 m. The derived declination of the source was not so easily obtained. Jansky used the fact that objects at different declinations trace out different loci of azimuths as the Earth turns. The declination of the cosmic source of radio energy appeared to be $-10° \pm 30°$. He felt that it might be significant that it could be the same direction as the Sun's motion through space, but he later concluded that the source of the radio waves was the plane of the Milky Way galaxy, indeed even the very center of the galaxy. Jansky's results were first presented in April of 1933 at a meeting of the International Union for Radio Science. Jansky's discovery was subsequently front page news in the *New York Times*.

Over the next several years noted astronomers, particularly at Harvard, became aware of Jansky's work, but none of them picked up the torch at that time. Jansky himself was encouraged to concentrate on more practical matters for the phone company. His only immediate disciple was Grote Reber, a radio engineer by day in Chicago, who built a parabolic 31′ 5″ diameter radio telescope in his backyard in Wheaton, Illinois. Completed in 1937, Reber's telescope was used over the next 10 years for observations of meteor trains and for the first surveys of the sky at radio wavelengths. It was mounted to pivot in elevation, while being fixed on the celestial meridian. The declination was chosen by tilting the telescope up or down, and the objects were allowed to drift through the field of view as they transited the meridian. Reber's early efforts to detect cosmic radio waves at wavelengths of 9 cm and 33 cm were unsuccessful, but one interesting conclusion could be made. Under the assumption that the radio waves were due to black body radiation, the expected signals at 9 cm should have been thousands of times stronger than those detected by Jansky at his wavelength of 14.5 m. But none were observed by Reber, implying that another mechanism had to be found to account for the existence and production of radio waves. Reber did manage to confirm Jansky's detection of cosmic radio waves at wavelengths of 187 cm and $62\frac{1}{2}$ cm. His maps at those wavelengths with respective resolutions of 12° and 4° were not improved on for a decade.[95] To all true astronomical amateurs (i.e. lovers of the field) Grote Reber remains an inspiration and a hero.

The first review of radio astronomical discoveries was published in 1947 by Reber and Jesse Greenstein.[96] By that time

> [r]egions of emission other than the galactic center were found numerous. Surveys had been made over a wide range of frequencies. A rapidly fluctuating signal was found from a small region in Cygnus, difficult to explain as refraction in an electron cloud (now interpreted as scintillation from a 'point' source). Theories reviewed were found generally unsatisfactory, because of the high brightness temperatures of the sources. The solar phenomena then known included intense noise storms, circular polarization and possible gyromagnetic radiation from sunspots. An addendum [Greenstein] wrote four months later mentions flare-associated impulsive bursts, the detection of the 'quiet' sun, the optical thickness of the corona at low frequencies and the desirability of a search for hyperfine emission of hydrogen. Data had far outrun interpretation, and little we reported had involved any cooperation with astronomers. The former radar wizards did their own interpretation, starting careers in England and Australia as a new breed, radiophysicists. In the United States radio telescope building began five years later at sites near, or in connection with, optical astronomers, e.g., Harvard, Michigan and Cornell. At first the natural academic links had been with electrical engineering groups such as the Naval Research Laboratory, Ohio State, and Cornell, often supported from military funds. A major impetus came from the Netherlands, where prediction of the 21 cm line by H. C. van de Hulst (1945) combined with the strong interest of J. H. Oort in galactic structure and rotation ... International organization started with URSI [the International Union for Radio Science], as a meeting place for electrical engineers. Later the IAU [International Astronomical Union] provided a home and a Commission, and astronomical journals actively solicited radio astronomy papers.[97]

I am getting ahead of the story here. Suffice it to say that by the late 1940s theoretical interpretations of cosmic radio waves had been made by a number of astronomers in America, Western Europe, Australia, and the

Soviet Union. It was known that radio waves were not primarily produced by black body (thermal) sources. While Reber had known how to calibrate his data, the derived temperatures of a million degrees made him feel certain that they would not be believed by astronomers. By the late 1940s astronomers came to believe that other, non-thermal processes were important in the universe, such as synchrotron radiation, which is emitted by electrons accelerated by magnetic fields.

Big time radio astronomy grew out of the wartime radar engineering groups using leftover equipment once used by the Australians, British, Americans, or confiscated from the Germans. Already during the Second World War, J. S. Hey and his group in Britain had discovered that sunspots emitted radio waves. This was not made public until after the war for the obvious reason that the Germans could carry out air raids at times of solar outbursts, when British radar would not be able to see them coming owing to the solar radio interference. The Moon was measured to have a temperature of 290 K at a wavelength of 1.25 cm by R. H. Dicke in the United States. This result was also made known after the war, by which time motivated teams of radar engineers who had done so much during the war were now looking for peacetime activities that could benefit from the same administrative structure. The efforts in Australia under J. L. Pawsey,[98] and in Britain under Martin Ryle at Cambridge,[99] and under Bernard Lovell at Manchester,[100] showed how much could be accomplished by an organized team under the direction of an effective manager who was also a top-notch scientist.

Among the instrumental highlights were the construction of the Parkes 210 foot (64 m) steerable radio telescope in Australia (1961), the Jodrell Bank 250 foot (76 m) Mark I steerable radio telescope (1957), and the development of aperture synthesis methods at Cambridge. The last-mentioned is particularly important because it allowed determinations of positions of radio sources with an accuracy rivalling optical astronomy. The resolution barrier of a telescope is fixed by the ratio of the wavelength of light at which it operates to the diameter of the telescope. For a 6 inch optical telescope this allows details better than 1 arc second to be seen. Because radio wavelengths are so much longer, a correspondingly larger radio telescope would be required to give the same resolution. A shortcut is to link two radio telescopes by cable or to record the signals received by two or more radio telescopes along with highly accurate atomic clock signals, thus producing a telescope whose effective diameter is the maximum separation of the individual elements, up to the diameter of the Earth. (Of course, light-gathering power is still measured by the total collecting area of the telescope(s).) Obtaining data by the method just mentioned is called very long baseline interferometry (VLBI). One must not only be adept in the areas of electronics and astrophysics, but one must also rely on the mathematics of Fourier Transform theory. Such a VLBI system gives interferograms, which are then transformed by digital computer into spatial maps of radio sources.[101]

With improved sensitivity and directability of antennas astronomers proceeded to make more and more detailed maps of the radio sky. The first Cambridge catalogue (1950) contained 50 individual sources. The second (1955) contained 1936 apparent sources. However, maps of overlapping areas of the sky by the Australians had little in common. The Cambridge group had overestimated the number of sources near their detection limit because of the 'confusion' of multiple sources in their highly elongated beam pattern.[102] The revised third Cambridge catalogue[103] (1962) contains only one-quarter as many sources as its Cambridge predecessor and has been used as a standard reference ever since.

Dutch radio astronomy[104] was motivated by the pioneer of modern galactic astronomy Jan Oort. During the war his colleague H. C. van de Hulst had calculated that neutral hydrogen should be detectable at a wavelength of 21 cm. While this detection was first achieved at Harvard by E. M. Purcell and H. I. Ewen in March of 1951, the Dutch and Australians were not far behind. The Dutch followed up with surveys of the sky at a wavelength of 21 cm, constructed a 25 m steerable dish at Dwingeloo (1956), and built the Westerbork Synthesis Radio Telescope (1970), consisting of 12 equatorially mounted 25 m parabolic dishes which can be placed over a baseline 1.62 km long.

If the first phase of radio astronomy contained only Jansky and Reber, the second concluded with the exploitation of the first national or academic centers for the initial surveys at radio wavelengths. By then the positions of some radio sources were known accurately enough to allow optical identification. Virgo A and Cygnus A were found to be galaxies at great distances – this contradicted the then-popular idea that most radio sources were nearby unusual radio stars that gave off radio, but not optical light. The identification of radio sources with the 48 inch Schmidt and the 200 inch Hale telescope at Palomar led in 1955 to the creation of Caltech's Owens Valley Radio Observatory, operational since 1958. Another class of objects then known at radio and optical wavelengths was supernova remnants, the first two examples being Taurus A (the Crab Nebula, remnant of the supernova seen in 1054) and Cassiopeia A (remnant of a star that blew up about 1660 but which was not seen due to strong interstellar absorption).

It became evident that radio sources associated with galaxies were often double. Cygnus A was the first such double source found. It is between the two radio lobes that one finds the galaxy seen at optical wavelengths.

Planets also fell into the realm of study of radio astronomers. In 1955 B. F. Burke and K. L. Franklin of the Department of Terrestrial Magnetism of the Carnegie Institution of Washington serendipitously discovered that Jupiter gives off bursts of radio waves. In 1956 scientists at the US Naval Research Laboratory measured Venus at a wavelength of 3 cm and found that the veiled planet had a surface temperature in excess of 600 K. That same group, under the direction of C. H. Mayer, measured the thermal emission from Mars and Jupiter in 1956. Other radio astronomers detected thermal emission from Saturn (1957), Mercury (1962), Uranus (1964) and Neptune (1966).[105]

Third phase of radio astronomy would include the construction of such centers as the Westerbork Radio Synthesis Telescope, improvements at Jodrell Bank, and the construction of facilities such as the National Radio Astronomy Observatory, the 1000 foot telescope at Arecibo, Puerto Rico, and the RATAN-600 in the Soviet Union (where radio astronomy had been carried out as long as anywhere else).[106]

As we have seen in a previous section, the time was ripe in the 1950s for the establishment of an American national optical observatory, and the same conditions led to the establishment of a national radio observatory.[107] The National Radio Astronomy Observatory (NRAO) came to be located in Green Bank, West Virginia. It is funded by the National Science Foundation and administered by Associated Universities, Inc., a research consortium run by nine American universities: Columbia, Cornell, Harvard, Johns Hopkins, MIT, Pennsylvania, Rochester and Yale. (This is not to be confused with the different research consortium, the Association of Universities for Research in Astronomy – AURA – that administers the National Optical Astronomy Observatories: Kitt Peak, Sacramento Peak and Cerro Tololo Inter-American Observatory.)

NRAO got its start in 1954 and had its first operational radio telescope in 1959, the 85 foot Howard Tatel telescope, which can now be linked with two other moveable 85 foot telescopes, forming an interferometric array that can give a maximum baseline of 2700 m (8900 feet). A transportable 45-foot telescope can also be added to augment the system. The dishes are made accurate enough to be used at wavelengths as short as 2 cm. This array was the prototype of the Very Large Array (VLA).

NRAO also has a 300 foot transit telescope (1962; adjustable in declination but fixed in the north–south meridian) and a 140 foot fully steerable dish (1965). At Kitt Peak they operate a 36 foot telescope, the surface of which is accurate enough to allow millimeter wavelength receivers to be used. Why is one of the NRAO telescopes at Kitt Peak while the others are located

Fig. 8.26. The 1000-foot Arecibo radio telescope. The Arecibo Observatory is part of the National Astronomy and Ionosphere Center, which is operated by Cornell University under contract with the National Science Foundation.

essentially at sea level? It depends on the wavelength of light one is interested in. At optical, infrared, and sub-millimeter wavelengths, the higher one is above sea level, the better the atmospheric transmission. But at wavelengths beyond 1 cm the Earth's atmosphere is transparent, even if it is raining, so altitude is not important.

The annual budget at NRAO has increased from its original $359 000 to the present $15 million, which includes the operation of the VLA in New Mexico.

The largest radio telescope, as measured by collecting area, is the 1000 foot (305 m) telescope at Arecibo, Puerto Rico (latitude $18\frac{1}{3}$ °N), which also serves as a radar dish for conducting experiments *on* the Earth's ionosphere and for carrying out radar mapping of various objects in the solar system. It was erected in 1960–3 by Cornell University under contract to the Department of Defense. In 1969 its operation was taken over by the National Science Foundation, making it part of the National Astronomy and Ionosphere Center. The telescope is built in a natural valley and is a fixed spherical dish. Different parts of the sky can be observed at the same time because a spherical collector does not bring all the light to a single focus. By varying the position of the moveable, suspended receivers, declinations from $-2°$ to $+38°$ can be reached. This allows all the planets and many asteroids to be observed.

How the Arecibo telescope came to be the world's largest is an interesting story.[108] The goal was to use radar and measure the signal reflected from electrons in the ionosphere to make 'laboratory' tests. Using accepted physics ideas it was calculated that the signals would be just detectable with a 1000 foot reflector and the radar transmitters in existence at that time. The large projected cost ($9 million) was declared acceptable by the Department of Defense when it was found that radar could be used to detect the ionization trails behind artificial satellites. Here was another technological challenge brought about by the launch of Russia's first Sputnik in October of 1957. When the Arecibo dish was nearing completion in 1962 a physicist named Kenneth Bowles noted that the accepted ideas of the ionosphere had been a bit naive. The motions of electrons in the ionosphere would be strongly influenced by the presence of charged atomic nuclei. The expected radar signal would be 100 times narrower than previously thought. It could therefore be detected with a telescope with 1% the collecting area. It could have been a 100 foot instead of a 1000 foot! Such telescopes already existed at that time, making the Arecibo dish 'superfluous'. However, this mistake has more than paid off if one considers the scientific dividends provided by the much larger dish. Astronomers can always think of ways to take advantage of an increase of light-gathering power!

The original dish was designed to be operated at wavelengths longer than 50 cm. Therefore, a wire mesh surface was all that was required. The old surface was replaced in the early 1970s with 38 778 adjustable aluminum panels, accurate enough to allow operation at wavelengths as short as 6 cm. Present (1986) plans for the telescope include: enlarging it to a diameter of 1200 feet; replacing the suspended receivers by a concave reflecting surface beyond the prime focus (a Gregorian reflector system) so that the receivers can be placed at the much more convenient ground level; and making the present support system interactive rather than the present passive arrangement.[109]

Because of the upgrades at Arecibo, which have amounted to more than the original construction cost, the telescope has far outlived its original projected life of 7 years. Its radar capability has been used to pulse the ionosphere and investigate the electrons and ions there. Radar mapping has been carried out on Venus and Mercury. In fact, it was with the Arecibo dish that the 59 day rotational period of Mercury was discovered.[110] It has been used in very long baseline interferometry experiments. Pulsars have been studied and discovered there. One of the public relations projects of the Arecibo dish involved beaming a coded message in the direction of the globular cluster M13 on 16 November 1974. While it will take tens of thousands of years for the message to reach potential inhabitants there, their equipment (expected to be much more advanced) might just receive our message. The decoded message translates into a rectangular diagram indicating the size of the telescope and the size of humans in terms of the wavelength of light used to transmit the message. There is information about where the message came from and that its senders knew the building blocks of deoxyribonucleic acid (DNA).

The reader will recall that optical astronomers of the nineteenth century undertook larger and larger scale surveys of the sky to measure star positions, obtain spectral types, measure parallaxes and proper motions. They pushed to obtain quality data on as many stars as possible, which is to say that they aimed for accurate measurements of fainter and fainter objects. A similar situation arose in radio astronomy. Astronomers sought to map the sky at key wavelengths such as 21 cm (to obtain the distribution of neutral hydrogen gas); they sought to investigate the composition of interstellar space and succeeded in finding an ever greater variety of molecules there (OH, water, ammonia, formaldehyde, ethyl alcohol, hydrogen cyanide, etc.); they sought to measure fainter and fainter objects; they endeavored to make high resolution maps of unusual sources to help unravel their nature; they discovered and monitored radio sources whose light output was variable over short time scales; data were obtained on the rotation of galaxies which have been important for the calibration of the expansion rate of the universe.

One of the most significant discoveries of the 1960s was achieved by Jocelyn Bell and Anthony Hewish at the Mullard Radio Astronomy Observatory of Cambridge University.[111] In an attempt to investigate the scintillation (twinkling) of radio sources, Bell (now Bell Burnell) noted some unusual traces on the strip chart records from their $4\frac{1}{2}$ acre aerial. When the source CP 1919 was subsequently observed in late 1967 by sampling the signal with much higher time resolution, it was found to pulse on and off with a period of 1.34 seconds. This implied either signals by Little Green Men or a physical source that was very small by astronomical standards. (Since information cannot be transmitted faster than the speed of light, an upper limit to the physical size of a source is obtained from the product of the velocity of light and the period of observed light variations.) It soon became evident that pulsars, of which hundreds are now known, had to be rapidly

Fig. 8.27. The Very Large Array near Socorro, New Mexico. Photograph courtesy of the National Radio Astronomy Observatory, operated by Associated Universities, Inc. under contract with the National Science Foundation.

rotating neutron stars each a few miles in diameter. It is noteworthy that the 1974 Nobel Prize in physics was shared by two radio astronomers – Tony Hewish and Martin Ryle.

Recent developments in radio telescope technology have included the construction of large fully steerable dishes and more advanced arrays. The world's largest fully steerable radio telescope, completed in 1972 and mounted in altazimuth fashion, is to be found at Effelsberg, West Germany.[112] It is run by the Max Planck Institute for Radio Astronomy in Bonn. It is a 100 m (328 foot) dish with a surface accurate enough to allow observations at wavelengths as short as 1.2 cm. The advantage of a large dish is its light-gathering power (proportional to its collecting area). Having it fully steerable makes for much less complicated data acquisition and analysis. While arrays of radio telescopes are good for making maps of objects, a large single dish is better for line spectroscopy of radio sources.

The most sophisticated radio astronomy installation in existence is the Very Large Array, located on the plains of San Augustin near Socorro, New Mexico.[113] It consists of 27 25 m (82 foot) parabolic dishes that can be arranged in a Y-shaped pattern over an extent of 21 km along the east or west arms, and 19 km along the north arm. Observations were made with the first antenna to arrive in October 1975, and the complete telescope was dedicated 5 years later. Its overall cost of construction was almost $80 million. The effective collecting area is equal to that of a 130 m reflector (one-sixth that of the Arecibo dish), but because it can operate effectively as a 27 km diameter telescope, it can achieve an angular resolution equal to that of the best ground based optical telescopes. This allows optical, infrared and radio images of nebulae and galaxies to be easily compared.

One of the most remarkable developments in radio astronomy has been the use of VLBI with continent-sized baselines to make maps of radio sources at a resolution of a thousandth of a second of arc – far better than any method of astronomy at any wavelength. A compact radio source has been found at the center of our galaxy which may be a supermassive black hole.[114] Maps of quasars have been made accurately enough to allow their rates of expansion to be measured. This has led to the discovery of the so-called superluminal radio sources – sources that appear to be expanding at several times the speed of light.[115] The first observations of this kind (1967) were made by David S. Robertson in Australia and A. T. Moffet at the Owens Valley Radio Observatory in California. Just about this time (1970) Irwin Shapiro (then at MIT, now the Director of the Harvard–Smithsonian Center for Astrophysics) and his colleagues used a radio telescope in southern California and one in Massachusetts to determine that 3C 279 was apparently expanding at 10 times the speed of light. The Shapiro group is usually given the credit for discovering superluminal sources because their independently obtained results appeared in print first and in far greater detail.

Another major discovery in the realm of radio astronomy was the discovery of the 3 K background emission by Arno Penzias and Robert Wilson in 1965,[116] for which they were awarded the 1978 Nobel Prize in Physics. Using a very accurately calibrated horn antenna at Bell Labs in New Jersey, where radio astronomy began, Penzias and Wilson found at a wavelength of 7.3 cm that even 'blank' sky gives off radio waves. This is interpreted as evidence of the Big Bang 10–20 billion years ago. The energy from that primeval explosion has become diluted because of the general expansion of the universe, such that it now corresponds to a black body at 3 K. While Penzias and Wilson first demonstrated the existence of this radiation beyond any reasonable doubt (their value was 3.5 ± 1.0 K), there had already been earlier evidence for it. In Japan H. Tanaka and his colleagues had found evidence in 1951 for background radiation corresponding to a temperature of 0–5 K.[117] In Canada, Medd and Covington had found evidence in 1952 for background radiation corresponding to a temperature of 5.5 K but its uncertainty was ± 20 K.[118]

Surveys of the sky at radio wavelengths are no longer the bread and butter of radio astronomers. Most are busy using major analytical instruments such as the VLA which allow a great number of observing teams to carry out their projects without having to know all the details of the operation of the instrumentation. While this makes it easy on the observer in the short run, eventually we could have a large population of theoretically sound scientists who regard most of the equipment as black boxes. This type is often regarded with a certain measure of contempt by engineers, technicians and instrument builders.

In his book *Cosmic Discovery*, Martin Harwit notes that no new astrophysical phenomena (such as the background radiation, superluminal sources, pulsars, and quasars) have been discovered at national centers.[119] Instead, one repeatedly hears names such as Caltech, MIT, Harvard, Cambridge and Bell Labs. With uncertain funding for many major projects, should we spend large sums on major national facilities so that everyone with a 'union card' at least can get a handful of nights of telescope time each year, or should we shift more funding to individual research groups that have demonstrated remarkable ability to produce results? A compromise is to allocate large amounts of telescope time to certain groups or projects at the national facilities so that experiments can be carried out to a greater depth than is presently allowed.

We have now covered much of ground-based astronomy at optical, infrared, and radio wavelengths – astronomy that *can* be done from the surface of the Earth. There is also major interest and activity in astronomy carried out from airplanes, balloons, rockets and satellites. What makes a tall mountain advantageous makes going higher even better. More new windows on the universe have been opened, from various infrared wavelengths to the whole gamut of high energy astronomy – ultraviolet, X-ray and gamma ray astronomy.

Airborne observatories and satellites

While ground-based astronomers can observe some of the time at certain far infrared and submillimeter wavelengths (30 microns to 1 mm), this can be done only on the tallest of mountains, and not very often. For some of these wavelengths no mountains are tall enough. Astronomy at wavelengths shorter than 3000 Ångströms *must* be done outside the Earth's atmosphere. The obvious solution is to have airborne observatories and satellites.

I now use the term 'observatory' in a limited sense. Since the beginning of the space age in October 1957 with the launch of Sputnik, we humans have successfully landed spacecraft on the Moon, Venus and Mars, and have conducted successful flybys of Mercury, Venus, Mars, Jupiter, Saturn, Uranus and Halley's Comet. While these space probes have added greatly to our knowledge of these solar system bodies, to the point of actually sampling non-terrestrial soil (and in the case of the manned Apollo missions, bringing it back to Earth), these were not general purpose observatories designed to survey the whole universe. Their respective missions were usually to carry out intensive study of one specific body. With my apologies to the Jet Propulsion Laboratory and the makers of orbiting solar observatories, I will reluctantly not cover this area of solar system astronomy.[120] Similarly, because rockets and balloons allow scientists to gather data for comparatively short periods of time, I shall only give them passing mention.

The use of relatively conventional jet airplanes to carry out astronomical pursuits in the 30 micron to 1 mm regime dates from 1966 when Frank Low and Carl Gillespie flew 14 flights on a Douglas A3-B bomber and made accurate measurements of the brightness temperature of the Sun using a 1 mm wavelength radiometer and then a helium cooled germanium bolometer.[121] The following year, under the banner of Gerard Kuiper, a Convair 990 aircraft at NASA Ames Research Center, Moffett Field, California, was used as a platform for the first airborne spectroscopic observations of the planets. This was the Galileo aircraft, which continued airborne missions, primarily for Earth-oriented experiments, until it crashed in April of 1973, with a loss of many lives. Its successor, the Galileo II, carried on with such projects as remote sensing of Mayan ruins in the Central American jungles. It burned up on the runway at Edwards Air Force Base in July of 1985, luckily without any human injuries.

The first astronomical observations with the Lear Jet Observatory at NASA Ames took place in October 1968. It carries a 12 inch Cassegrain telescope to altitudes as high as 50 000 feet.

How can one operate a telescope in an airplane given that the Earth is turning, the plane is moving, and the ride is not completely smooth? The telescope looks out of one side of the plane, mounted on some kind of air bearing, and preferably stabilized by gyroscopes. The plane must fly various curved paths over the Earth to keep the object of interest at the same relative azimuth with respect to the plane.

It had to be demonstrated first that a telescope port is compatible with the aerodynamics of an aircraft. It was found that if a spoiler is placed just fore of the telescope so that air can flow over the telescope opening, local turbulence in the field of view was minimized and acceptable images were the result. Most importantly, the airplane flew just fine.

The Lear Jet Observatory (LJO) was the prototype for the much larger

Fig. 8.28. The Kuiper Airborne Observatory, which carries a 36 inch reflecting telescope to altitudes as high as 45 000 feet. The telescope is situated in a cavity located just fore of the left wing. NASA photograph.

Kuiper Airborne Observatory. Because the Lear Jet is much smaller, it is relatively inexpensive to operate, and hence has been used extensively for the development of instrumentation. Research carried out on the Lear Jet has involved photometry and spectroscopy of planets, stars, and nebulae. It has recently been upgraded, at a cost of $1 million. It is particularly suited for projects such as comet observations because the telescope can view all the way down to the horizon.[122]

The Kuiper Airborne Observatory (KAO; Fig. 8.28), which carries a folded Cassegrain telescope to altitudes as high as 45 000 feet, is a modified C-141 4-engine military transport jet.[123] It began operations in 1974. The intervening years have shown it to be a unique platform for astronomical research. Data flights on the KAO (typically $7\frac{1}{2}$ hours) are much longer than those of the LJO ($2\frac{1}{2}$ hours). Actual observing time is $6\frac{1}{2}$ hours per flight on the KAO vs only 70 minutes per flight on the Lear jet. At an effective cost of more than $50 000 per KAO flight, time is really money, so the pressure to take data effectively and efficiently is even more intense than at regular ground based observatories (where a typical night's observing on a 4 m telescope is worth $10 000). Of course one tries to check everything out on the ground, but the rule always seems to be: something always works differently once the wheels leave the ground. Electrical grounding is particularly crucial, and many a hardware problem is linked to ground loops. Because the airplane experiences some turbulence, the astronomical detectors must be designed to cope with sources of microphonic noise.

A typical flight has on board the flight crew of three, the experimenters (2–7), the mission director, telescope operator, tracking telescope operator and computer operator, each of which plays an active role in the operation. Because no repairmen can be beamed up to the plane, the crew must function like astronauts in space and be able to handle problems as they arise, fix them with help radioed from the ground, or choose to abort the flight.

The KAO typically flies 80 research flights per year, and because it is mobile, it can go wherever it needs to for the observations. While most flights begin and end at Moffett Field, California, expeditions go to Hawaii often so that the Galactic Center can be studied on long flight legs. Occulations of stars by planets or asteroids have been observed from far away places. In fact, it was from Australia in March of 1977 that the KAO made the first observations of the rings of Uranus. (Ground based observers also made this discovery simultaneously.) In July of 1981 a solar eclipse was observed with the KAO off the Kuril Islands north of Japan. Halley's Comet was observed from California in the fall of 1985 and from Australia in the spring of 1986.

Observations made on the KAO have spanned an extremely wide range. Planetary atmospheres have been extensively studied. Sulphuric acid and water vapor have been found on Venus. (While the casual reader may not be impressed with the detection of water vapor on other planets or in comets, remember that one must be able to eliminate the effect of terrestrial water vapor in the spectra. Unless one has an orbiting telescope, the calibration must be carefully done.) Jupiter has been one of the objects most studied by the KAO. While the prime constituents of its atmosphere (molecular hydrogen, helium, methane and ammonia) were discovered before there was airborne astronomy, many trace chemicals have been discovered on KAO flights. The list includes acetylene (C_2H_2), phosphine (PH_3), monodeuterated methane (CH_3D), hydrogen sulfide (H_2S), ethane (C_2H_6), hydrogen fluoride (HF) germane (GeH_4), and water vapor. Many of these species have been successfully found on Saturn.[124]

Measurements of the heat energies of the outer planets have been made on the LJO, the KAO, and the NASA Infrared Telescope Facility by the University of Chicago/Yerkes Observatory infrared group. They find that Jupiter, Saturn, and Neptune radiate 3.2 times as much energy as they receive from the Sun, indicating that each has an internal heat source. However, Uranus has a dead core, since it is only as hot as the feeble sunlight makes it (58° above absolute zero).[125] These thermal measurements are particularly important for the calibration of many astronomical experiments.

It has been said that an infrared astronomer could make a good living observing only the Orion Nebula and the Galactic Center.[126] The Orion Nebula certainly has a lot of noteworthy features: protostars, ionized and neutral gas, a molecular cloud, and shock fronts. The Galactic Center is the only galactic nucleus we can study up close, and because of strong interstellar absorption by dust it is best studied at infrared and radio wavelengths. The KAO has played a key role in the development of theories of regions of star formation by allowing observations of spectral lines of various atoms in different states of ionization and of molecules at different states of vibration and rotation. Broad band photometry of protostars at various wavelengths inaccessible from the ground has also been very useful. Because a star's mass governs its subsequent evolution more than any other parameter, and a star's luminosity is a strong function of its mass, observations of protostars of different luminosities allows us to investigate any differences in the formation rates of low mass and high mass stars. Since half of the mass in the local portion of the galaxy is still unidentified, there remains much to be understood concerning how our galaxy came to be what it is today.[127]

The Americans have not been the only ones involved in airborne astronomy. Astronomers at the Observatoire de Meudon have flown a 32 cm telescope on a French Caravelle and the NASA Convair 990. Balloon borne infrared telescopes have been operated by universities and research institutions in the United States, United Kingdom, India, Japan, and West Germany.

Of particular importance in high altitude infrared astronomy is the Air Force Geophysical Laboratory survey made in the mid-1970s. It covered 90% of the sky at wavelengths of 4, 11, 20, and 27 microns. This was done on 11 rocket flights, each with only 5 minutes of observing time. This would have taken a year's *observing* time with a ground based 60 inch telescope. The AFGL survey used a 16.5 cm telescope. The limiting sensitivity at 11 microns was a flux of 100 Janskys.[128]

The most complete survey of the infrared sky has been carried out by the Infrared Astronomical Satellite (IRAS) a joint project by the United States, United Kingdom, and The Netherlands.[129] Launched in January of 1983, IRAS was a 57 cm (22.4 inch) telescope cooled to a temperature of $2\frac{1}{2}$ K. (This was so that the telescope, which would otherwise emit its own infrared radiation, did not see itself.) IRAS' 72 kg of superfluid helium coolant ran out in November of 1983, but not before IRAS had surveyed 96% of the sky twice. It was three-quarters of the way through its third pass.

IRAS operated at wavelengths of 12, 25, 60 and 100 microns for its survey work. It also had a low resolution spectrometer that gave spectra from 8 to 22 microns. While the data analysis will be carried on for years to come, the most significant results have been known for some time. IRAS discovered a total of six comets, infrared 'cirrus' clouds in the galaxy, and a cloud of material surrounding Vega which is like a system of planets or asteroids in formation. The IRAS catalogue[130] contains 245 839 point sources and 16 740 extended sources. There are 5425 8–22 micron spectra. The flux limit at 12, 25 and 60 microns was 0.5 Jansky – 200 times better than the AFGL survey. While most of the point sources will turn out to be relatively normal stars and galaxies, a small fraction is already known to have no known counterparts at other wavelengths. It is objects such as these that pave the way for many new scientific insights.

The first dedicated ultraviolet satellite was the Orbiting Astronomical Observatory number 2 (OAO-2) launched in December 1968. It consisted of eleven telescopes: four $12\frac{1}{2}$ inch telescopes and four 8 inch telescopes, all used for broad band photometry of stars from 1050 to 3200 Å by means of television-like cameras; a 16 inch telescope for photometry of nebulae from 2100 to 3400 Å; and two 8 inch telescopes for taking spectra from 1000 to 2000 Å and from 2000 to 4000 Å, respectively.[131] The mission of OAO-2 was to carry out the first extensive ultraviolet survey by measuring the ultraviolet brightnesses of as many stars as possible, and to obtain spectra of selected sources. It is a rule of thumb that the most accurate way to measure the colors and luminosities of stars is to observe them near the wavelengths where they are brightest. These first ultraviolet observations were the first such accurate measurements of many known hot stars.

A successor was OAO-3, launched in August 1972. It was subsequently known as Copernicus in honor of the 500th birthday of the famous Polish astronomer (who was born in 1473). The Copernicus satellite consisted of a 32 inch reflecting telescope for taking spectra over the range 700 to 3200 Å, three small X-ray telescopes that operated at wavelengths on the order of 70 Å, and a proportional counter which detected photons in the 1–3 Å range.[132] The 32 inch ultraviolet telescope on OAO-3 was operated 90% of the time. Because of its much greater light-gathering power, Copernicus could take spectra of much fainter stars. It operated for 9 years until it was turned off in 1981.

The most productive ultraviolet satellite to date has been the International

Ultraviolet Explorer (IUE) a joint project by the United States, United Kingdom, and the European Space Agency (ESA).[133] Launched in January 1978, it is still operational at the time of this writing. IUE is a 45 cm (17.7 inch) reflecting telescope designed entirely for spectroscopy. While it is smaller than Copernicus, improved detector technology has allowed it to see fainter stars (to 16th magnitude). It orbits high above the Earth (perigee: 25 700 km; apogee: 45 900 km) so that it is always visible from the United States or Europe. For 16 hours a day it is operated from the Goddard Space Flight Center near Washington, DC. The other 8 hours each day it is operated by ESA at a ground station near Madrid. IUE can take spectra either from 1150 to 1990 Å, or from 1850 to 3200 Å. Instead of one very long, narrow spectrum, each IUE 'image' is a stack of spectra like lines of music paper with spectral lines instead of notes running left to right on each staff.

Ultraviolet astronomy, from 1000 to 3500 Å, is the astronomy of objects hotter than the Sun's photosphere, typically 10 000 to 1 million K. This includes massive early-type stars like those lighting up galactic nebulae, the Sun's chromosphere, other stars with active chromospheres, stars that are covered by large star spot areas (just like sunspots, only taking up a larger portion of the stellar surface), and interstellar gas heated by very hot stars. The extreme ultraviolet region from 100 to 1000 Å has only recently been opened. Only a couple of dozen EUV sources are known, but NASA will be launching an extreme ultraviolet satellite in the late 1980s.[134]

The X-ray region of the electromagnetic spectrum ranges from wavelengths of 0.1 to 100 Å (photon energies of about 100 KeV to 0.1 KeV). Celestial X-rays are given off by sources undergoing even higher energy processes than those that only give off UV and visible light. While novae, supernovae, spotted stars, and the Sun's corona are all X-ray sources, the most interesting discoveries in the X-ray world are objects that may very well turn out to be black holes.

The first X-ray observations were carried out with Geiger counters carried to high altitudes by rockets such as the V-2s and Aerobees. The Sun was the first X-ray source identified in the sky. The key group that carried on this work in the 1940's and 1950s was based at the Naval Research Laboratory in Washington, DC.

The key group responsible for much of the initial successes in extra-solar system X-ray astronomy was that at a private company called American Science and Engineering (AS&E), founded in 1958 in Cambridge, Massachusetts. Riccardo Giacconi was the leader of what became a highly motivated and innovative research team. (In our survey of astronomical centers and the associated 'royalty', Giacconi must be regarded as the one time czar of X-ray astronomy. He is now director of the Space Telescope Science Institute.) Like the British and Australian radio astronomy groups after the Second World War, Giacconi's group at AS&E showed what teamwork and effective management could do.

In 1961, in response to testing of nuclear weapons by the Soviets, President Kennedy decided to carry out similar tests in the United States and to measure the effects of bursts of nuclear weapons at high altitudes. AS&E was asked to produce payloads to measure the particles produced by the explosions. Their staff increased by an order of magnitude to meet the demand. The knowhow to study the effects of man-made high energy processes led directly to the desire to measure such processes originating in the cosmic realms.

In the summer of 1962 a rocket was launched by the AS&E group from White Sands, New Mexico. The scientific goal was to detect X-rays from the Moon, but instead a source was discovered in the constellation Scorpius, known since as Sco X-1. This discovery was confirmed soon by the NRL group under Herbert Friedman, which also found the Crab Nebula to be an X-ray source.[135]

Rocket flights during the 1960s brought the number of known X-ray sources to about 30. The first all sky survey at X-ray wavelengths was carried out by the Small Astronomical Satellite number 1, an American satellite launched from Kenya in December 1970, and known as Uhuru after the Swahili word for freedom. Analysis of the Uhuru data eventually revealed 339 X-ray sources. Most of these are distributed along the plane of the Milky Way, indicating that they are in our galaxy. Such sources include supernova remnants, X-ray novae, X-ray binary stars, and X-ray pulsars. Extragalactic sources such as the Magellanic Clouds, the radio galaxy Cygnus A, the quasar 3C 273, Seyfert galaxies, and clusters of galaxies were also found to emit X-rays. Most of these sources are found to have variable X-ray flux. The most significant discovery of Uhuru was the source Cygnus X-1, an X-ray binary consisting of a 20 solar mass O-type star and an unseen companion whose mass is about 8.5 solar masses; this is much greater than the allowed mass for a neutron star, indicating that it may very well be a black hole.[136]

Uhuru and other X-ray survey satellites of that era launched by the Americans and Europeans did not have direct imaging capabilities. They could only determine the position of X-ray sources within one degree, leaving it up to specialized (non-survey) experiments at X-ray, optical, and radio wavelengths to help identify the actual X-ray sources. The first imaging X-ray telescope was the High Energy Astronomy Observatory number 2,

Fig. 8.29. The Einstein X-ray satellite. NASA photograph.

renamed Einstein in honor of the 100th birthday of that famous scientist.[137] It was launched in November 1978 and operated until April 1981. It was the second of three High Energy Astronomy Observatories launched by the Americans. Each focused the incident X-rays by means of a set of collimated parabolic, then hyperbolic, cylinders known as Wolter grazing incidence mirrors. One must have the X-rays graze off such mirrors because they would otherwise be absorbed by the metallic surface. The whole apparatus looked like a set of nested metal garbage cans, but the set of four mirrors in Einstein, for example, was figured more accurately than a conventional optical telescope mirror.

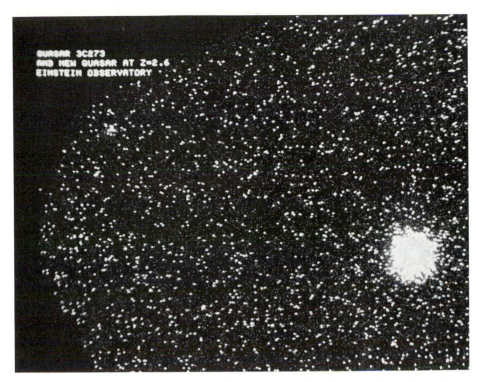

Fig. 8.30. Einstein satellite image of the quasar 3C 273. Photograph courtesy of J. Patrick Henry.

The Einstein Observatory had five detector systems: two spectrometers, a proportional counter, the High Resolution Imager (HRI), and the Image Proportional Counter (IPC). The IPC gave a field of view 75 arc minutes in size, with a resolution of 1 arc minute. It was very sensitive, being able to detect sources 1000 times fainter than Uhuru's limit. The HRI gave a 25 arc minute field of view with a spatial resolution of 2 arc seconds in the central 5 arc minutes of the field. This is accurate enough to allow easy identification of the X-ray sources for comparison with optical photographs.

In addition to studying exploding or exploded stars, hot intergalactic gas, galaxy clusters, and the like, the X-ray and ultraviolet satellites launched in the past 10 years have also discovered that red dwarf stars are copious producers of high energy radiation. The advent of ultraviolet and X-ray satellites has led to systematic monitoring of dwarf novae by dedicated amateur astronomers so that the satellites can be directed to sources at times of unusual activity.

The Americans have not been the only producers of high energy astronomy satellites.[138] In 1979 the Japanese launched the Hakucho ('Cygnus') X-ray satellite, which was joined in February of 1983 by Tenma

('Pegasus'). They operated until 1985. In March of 1983 the Soviets launched Astron, a non-imaging ultraviolet telescope with a 32 inch mirror. It operates in the 1100–3600 Å range. The orbit of Astron is highly elliptical, taking it halfway to the Moon. This allows for many hours of uninterrupted observations.

The successor to Einstein has been the European Space Agency's Exosat. Launched in May of 1983, it operated until April 1986. It had a pair of imaging telescopes and was comparable in sensitivity to Einstein, while operating at somewhat different wavelengths. Like Astron, Exosat was placed in a large, highly elliptical orbit that took it halfway to the Moon.

The highest energy light waves are gamma rays, with wavelengths less than about 0.1 Å (energies greater than about 0.1 MeV).[139] These photons have wavelengths considerably smaller than the spacing of atoms in solids, such that they do not interact with gas in a proportional counter or graze off of Wolter mirrors. Gamma-ray telescopes use two basic types of detectors. For energies less than 10 MeV one can use scintillation counters. Atoms are bombarded by the high energy rays and give off visible photons, which can then be detected with photomultiplier tubes. High energy gamma rays are detected by spark chambers, consisting of a stack of thin metal plates with spaces between them. The high energy gamma rays incident on the detector produce electron–positron pairs if the plates are operated at high voltage. (The positron is the antiparticle of the electron.) The electron–positron pairs produce ionization trails which are seen as sparks. By collecting the particles produced in the spark chambers and recording the cascades of the sparks, it is possible to determine the approximate energies and directions of the incoming gamma rays.

One of the principal problems of non-optical astronomy is being able to cope with extraneous signals – noise that is – due to the sky background or the instrument itself, rather than to the individual source under study. In the case of gamma rays, the telescope has to contend with a flux of cosmic ray particles that outnumbers the gamma ray flux by a factor of 100 000 to 1.

Gamma ray instruments have been operated on lunar landers and planetary flybys. The very highest energy gamma rays can be detected by ground-based Čerenkov detectors, but gamma ray astronomy is primarily based on observations from orbiting gamma ray satellites.

The two principal gamma ray satellite experiments carried out so far were done by the American Small Astronomical Satellite number 2, which operated for 7 months in 1972–3, and the European COS-B satellite, which operated from 1975–82. While only 250 000 gamma ray photons have been collected in all of gamma ray astronomy, some conclusions can be made about the nature of cosmic gamma ray sources. Solar flares have been detected. Pulsars and molecular clouds have also been detected. (While it may seem strange that generally cold clouds should emit gamma rays, it is so because high energy cosmic rays, say, from supernova explosions, magnetic fields around pulsars, or accretion disks around black holes, collide with atoms in the clouds, giving rise to high energy photons. It is the total amount of matter in the way that is important in this case, and molecular clouds are the most massive individual objects in the galaxy.)

One of the most intriguing gamma ray sources is called Geminga because it is a gamma ray source in the constellation Gemini. Discovered by SAS-2, Geminga's position was fixed by COS-B to within 0.5 degrees. There is an X-ray source within this error box, which was imaged by Einstein, but searches at optical and radio wavelengths have not revealed any good candidates for the X-ray source or Geminga, if in fact they are one and the same. No one is quite sure what Geminga is, and there are a number of other Unidentified Gamma-ray Objects (UGOs) to investigate. Perhaps they are so partisan to emitting gamma ray radiation that none have counterparts at other wavelengths. Once again, investigating a recently opened window on the universe has produced some surprises and many more questions. Much remains to be done to understand the variety of already known cosmic sources of electromagnetic radiation. Providing that there is adequate funding for presently planned projects, the late twentieth century will continue to be an era of great astronomical discovery.

9

The future

The true and lawful goal of the sciences is none other than this: that human life be endowed with new discoveries and power...
Francis Bacon, *Novum Organum* (1620).

Nature offers no greater splendor than the starry sky on a clear, dark night. Silent, timeless, jeweled with the constellations of ancient myth and legend, the night sky has inspired wonder throughout the ages. It is a wonder that leads our imagination far from the confines of Earth and the pace of present day, out into boundless space and cosmic time itself.
  Astronomy, born in response to that wonder, is sustained by two of the most fundamental traits of human nature: the need to explore, and the need to understand. Through the interplay of discovery, the aim of exploration, and analysis, the key to understanding, answers to questions about the Universe have been sought since the earliest times, for astronomy is the oldest of the sciences. Yet it has never been, since its beginnings, more vigorous or exciting than it is today.
  Astronomy Survey Committee quote in *Astronomy and Astrophysics for the 1980's.* 1 *Report of the Astronomy Survey Committee*, Washington, DC: National Academy Press, 1982, p. 3.

It is, of course, not enough to spend hundreds of millions of dollars on bigger, more powerful telescopes operating at a variety of different wavelengths. We must have specific research programs in mind. While astronomy has not given us too many immediate practical fruits, no one can deny that the discoveries of recent times have contributed greatly to the human equation. Seeking to understand our universe is as basic to human beings as breathing. Of the many questions we hope to answer in the near future are the following:

1) Are ice ages on the Earth caused by the collision of the Earth with comets or asteroids? Did comets kill off the dinosaurs?
2) How prevalent are planets in the galaxy?
3) Is there life on other planets?
4) Exactly how do stars form? So far we only know *that* stars form and where, but many of the details of the process are unknown.

5) In what form(s) is the unseen matter in the universe?
6) Do black holes really exist?
7) Similar to the previous question, what is at the core of our galaxy?
8) At what epoch of the evolution of the universe did different types of galaxies form? How do we account for the distribution of galaxies throughout the universe?
9) What are quasars? What powers them? How far away are they?
10) What is the expansion rate of the universe? Will the universe expand forever?

While there are an estimated 35 000 astronomers in the world today, and hundreds of professional caliber telescopes, the solution of the above-named questions (and many others of substantial merit) requires the ability to detect fainter objects with improved spectral resolution, spatial resolution and time resolution (the ability to measure the strengths and widths of more spectral lines, make more detailed maps of objects and to sample data faster). In this chapter we shall discuss a number of larger telescopes nearing completion or on the drawing board, both ground based telescopes and those that will operate above the Earth's atmosphere.

But first, the Multiple Mirror Telescope (MMT; Fig. 9.1). The future is now at Mt Hopkins, Arizona. The MMT, designed and built throughout the 1970s, was completed in the spring of 1979.[1,2] It consists of six 72 inch mirrors arranged in a hexagonal pattern to provide a single telescope equivalent to a 4.5 m diameter reflector, making it currently the world's third largest optical and infrared telescope. It was designed primarily for spectroscopic and photometric observations of individual objects in small sky fields, but its great light-gathering power has been exploited to obtain deep images of galaxies and quasars. Because of its trend-setting design, the MMT cost only 30% of a conventional single mirror 4.5 m telescope. Now that modern detectors have reached very high levels of performance, effectively increasing the number of photons measured by a factor of 30,* the next step is to construct telescopes as large as possible, but also as inexpensively as possible. The use of multiple mirror and segmented mirror designs, altazimuth mountings and computer-driven mirror support systems will allow for the best use of the advanced detectors available.[3]

Single and multiple 8 m telescopes are being planned by several groups. The Carnegie institution of Washington, which funds the Mt Wilson and Las Campanas Observatories, is planning to build an 8 m telescope at Las

---

* See footnote on p. 189.

Fig. 9.1. The Multiple Mirror Telescope, Mt Hopkins, Arizona. It has six 1.8 m
mirrors which are used to provide the light gathering power of a 4.5 m telescope.
Photograph courtesy of the Harvard–Smithsonian Center for Astrophysics.

Campanas, Chile. The University of Arizona and Johns Hopkins University
are collaborators in this project.[4] A consortium consisting of the University of
Arizona, Ohio State University, and the University of Chicago is planning to
build the world's largest 'binoculars' at Mt Graham, Arizona. They will
consist of two 8 m telescopes with a mirror center to mirror center separation
of 23 m.[5] The European Southern Observatory is planning the Very Large
Telescope (VLT), consisting of four 8 m telescopes.[6]

While the three just-mentioned telescopes have some funding hurdles in
the way before they can achieve reality, the W. M. Keck Observatory
(Fig. 9.2), to be sited at Mauna Kea, does not. Designed by University of
California astronomers over the past 10 years, it was originally going to be
the new big gun in the University of California arsenal, taking over from the
3 m Shane telescope at Lick Observatory. With the announcement in
January of 1985 that the W. M. Keck Foundation would donate $70 million

Fig. 9.2. The Keck Telescope, a 10 meter segmented mirror telescope to be sited
at Mauna Kea. It is planned to be operational by 1991. Photograph courtesy of
the California Association for Research in Astronomy.

to the California Institute of Technology for its construction, it became a joint
University of California–Caltech project. It will have a mirror consisting of 36
hexagonal segments each 1.8 m wide and only 7.5 cm thick. Though 10 m
in diameter, the whole mirror will be very light, thus simplifying the
construction of the altazimuth mounting. With its 4-fold increase in light
gathering power compared to the Palomar 5 m telescope, the Keck Telescope
will allow astronomers to see substantially further out into space and back
into time. Construction is scheduled for completion in the early 1990s.[7]

Also under construction at Mauna Kea and in fact nearing completion at
the time of this writing are two millimeter-wavelength telescopes, the 15 m
James Clerk Maxwell Telescope (JCMT; built by the British and Dutch) and
the 10.4 m Caltech Submillimeter Observatory. These will be used for the
investigation of the composition and energetics of the interstellar medium
and for uncovering whatever else the submillimeter and millimeter wavelength regimes of the electromagnetic spectrum have to offer. Until recently
submillimeter-wave astronomers have carried on the best they can with

Fig. 9.3. 'Millimeter-wave valley' inbetween the summit cone of Mauna Kea and Pu'u Poliahu, the site of the first telescope situated at the mountain. Photo taken December 1984. The octagonal building is for the 15 m James Clerk Maxwell Telescope. The doughnut-shaped foundation is for the 10.4 m Kresge Telescope being built by Caltech. These radio telescopes will operate in the submillimeter and millimeter wavelength ranges from 350 microns to 1.1 mm.
Photograph copyright by Michael Broyles, used by permission.

facilities such as the Kuiper Airborne Observatory and the United Kingdom Infrared Telescope, but the advent of the JCMT and the CSO at a high altitude site such as Mauna Kea will greatly enhance astronomers' ability to investigate a little studied spectral region by providing them with a great increase in light-gathering power.

The biggest American plans for ground-based astronomy involve the construction of a National New Technology Telescope (NNTT), a proposed 15 m telescope consisting of four 7.5 m telescopes in one (Fig. 9.4). It would cost upwards of $100 million and would operate from 3000 Å to 20 microns. Providing funding can be found, it is to be built at Mauna Kea. The NNTT would be operated like Kitt Peak, as a national facility administered by the National Optical Astronomy Observatories.[8]

The Soviets have also been concerned with new technology telescope design. Astronomers at the Crimean Astrophysical Observatory have been operating a 1.2 meter MMT since 1978. Called the AST-1200, it has seven

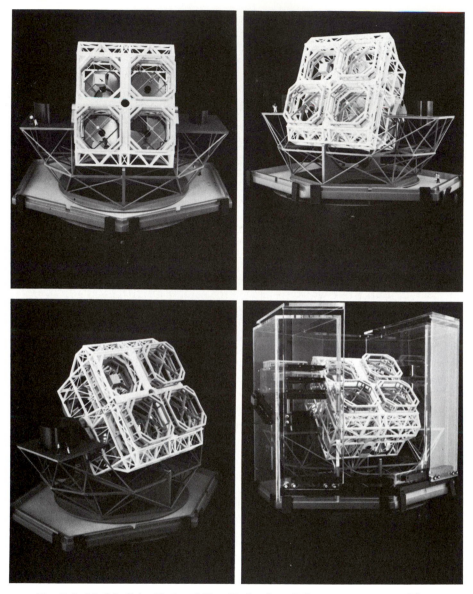

Fig. 9.4. Model of the National New Technology Telescope, to consist of four 7.5 m mirrors in a single instrument. Photograph courtesy of the National Optical Astronomy Observatories.

hexagonal mirrors each 0.4 m wide. The largest of all proposed ground based telescopes is the Soviet 3T-25, to have a 25 m mirror of about 500 hexagonal segments. It would have a 6 m diameter *secondary* mirror – equivalent in size to the world's presently largest telescope![9]

Fig. 9.5. Artist's conception of the Stratospheric Observatory for Infrared Astronomy (SOFIA), a hoped-for NASA facility which will have a 3 m telescope. NASA photograph.

In the realm of airborne astronomy astronomers have talked for some years about building a larger facility similar to the Kuiper Airborne Observatory. One such possibility is the Stratospheric Observatory for Infrared Astronomy (SOFIA, Fig. 9.5), which would fly a 3 m telescope on a Boeing 747-SP to altitudes as high as 45 000 feet, but for longer duration than KAO flights. With a light-gathering power ten times that of the KAO 36 inch telescope, SOFIA would make the difficult experiments much easier and open up new vistas for suborbital astronomy.

The most eagerly awaited event in the world of astronomy is the launch of the Hubble Space Telescope (HST), delayed at the time of this writing because of the explosion of the Space Shuttle Challenger on 28 January 1986. Hopefully, the Shuttle program will get back on its feet quickly with no more catastrophes and we may resume the bold course upon which we have embarked.

The Hubble Space Telescope[10] is a 2.4 meter telescope which will be the first telescope to utilize fully the possibilities of optical astronomy. It is the first general-purpose optical telescope to be operated above the Earth's atmosphere. It will allow observations from the ultraviolet at 1100 Å to the near infrared at 11 000 Å (1.1 microns). Its principal advantages will be its ability

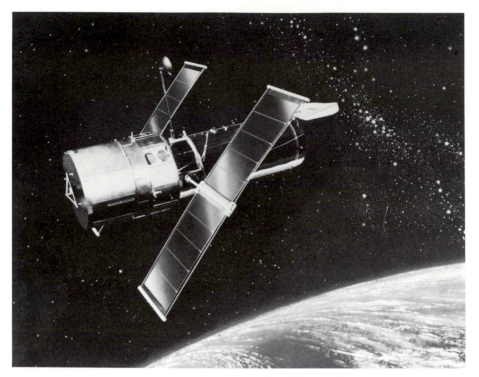

Fig. 9.6. The Hubble Space Telescope, which will be the first general purpose telescope in orbit to operate at optical wavelengths. It will also allow observations in the ultraviolet and the near infrared. NASA photograph.

to resolve details in cosmic sources (to within 0.05 arc seconds – ten times better than ground base sites) and its ability to detect fainter objects (to 28th magnitude). The brightness limit is roughly ten times better than that of a ground based 4 m telescope. The HST will be able to take advantage of the better contrast against the night sky because it is outside the Earth's atmosphere. This capability will allow users to probe much further out into space and back into time than will be possible even with the Keck 10 m telescope or the NNTT. We should at last get a firm handle on questions of cosmological significance. We will be able to determine the ultimate fate of the universe by investigating its very distant past. Will the universe expand forever, or will its expansion slow down and the universe eventually come together again in one Big Crunch? We hope to have an answer to this question by the end of the century.

The HST has a Fine Guidance System and five detector systems utilizing the most advanced solid state technology, image intensifiers, TV tubes and photomultipliers. The high resolution spectrograph allows UV spectra of objects brighter than 17th magnitude to be taken. The faint object spectrograph is for ultraviolet and optical spectra of objects fainter than 19th magnitude. The faint object camera is for ultraviolet and optical imaging in the 21–28 magnitude range over a field three-quarters the diameter of the full moon. The wide field planetary camera allows images of objects as bright as 9th magnitude over the full wavelength range of the HST. The high-speed photometer allows photometry of objects brighter than 24th magnitude and allows for a sampling rate as fast as 60 000 times per second. It can be used to measure stars as bright as Betelgeuse and also for monitoring the nuclei of faint galaxies and quasars almost too faint to be seen with the most powerful telescopes on Earth. Such range in one instrument is incredible!

The HST will be operated under some serious constraints. Because of the satellite's orbit, the Earth will occult objects every 90 minutes, so observations must be planned carefully in advanced. The telescope moves and sets slowly and will only be used about 3000 hours per year – comparable to that of a major ground-based telescope on a good site. Its biggest problem will be its popularity. The Space Telescope Science Institute in Baltimore, Maryland, expects there to be over 2000 observing proposals per year – 15 to 20 times the amount that can be accommodated.

Nearly all discoveries made at non-optical wavelengths have been explained or greatly elucidated at optical wavelengths.[11] Even with the imminent launch of new infrared and high energy astronomy observatories the HST should prove to be the most powerful tool available in the realm of observational astronomy. May its astronomical pricetag ($ 1.4 billion) be justified in terms of the results it will provide.

Two satellites are planned to follow up the Infrared Astronomical Satellite. They are the American SIRTF (Space – formerly Shuttle – Infrared Telescope Facility) and the European Space Agency's Infrared Space Observatory (ISO). ISO will have a 60 cm mirror and operate from 3 to 200 microns. SIRTF will have an 85 cm mirror (giving more than twice the light-gathering power of IRAS) and will operate from 2 to 700 microns. Each will be more than 1000 times more sensitive than IRAS, which itself was more than two orders of magnitude more sensitive than instruments used to carry out previous all sky surveys. It is expected that ISO and SIRTF will be launched in the early 1990s.[12] In the meantime, infrared astronomers are keeping busy analyzing the IRAS data and investigating IRAS sources with ground-based telescopes.

In the realm of high energy astrophysics there are three noteworthy projects to be mentioned. As this book goes to press the Germans plan to launch Rosat, named after Wilhelm Röntgen, the discoverer of X-rays. It will have an 83 cm diameter X-ray telescope – the largest one so far – and will provide a needed follow up to Exosat following that satellite's demise in April of 1986.

The Astronomy Survey Committee, the American group which sets the priorities for American astronomical funding each decade, stated in its last report issued in 1982 that the highest priority right now would be an Advanced X-ray Astronomy Facility. As yet it is unfunded, but it would cost on the order of $500 million. AXAF would have an effective aperture of 1.2 m. It would have 100 times the sensitivity and 10 times the angular resolution capability of the Einstein satellite. Its power would greatly increase our understanding of supernovae, galactic nuclei, and help in the search for black holes. Unlike other satellites, it is expected that AXAF would be serviced by Space Shuttle astronauts, allowing it to operate for 10 years or more.[13]

Gamma ray astronomy will come of age with the launch of the Gamma Ray Observatory (GRO) in the late 1980s. This NASA-funded probe, to operate over a range of 100 KeV to 30 GeV (wavelengths 0.1 to 0.0000004 Å), will bring gamma ray astronomy to the state that X-ray astronomy was in the early 1970s. Of particular importance to gamma ray astronomers will be the study of supernovae in galaxies as far away as the Virgo Cluster, the observation of gamma ray burst sources, pulsars, active galactic nuclei and the interaction of cosmic rays with the interstellar medium – all made possible by the launch of the GRO. The elucidation of the nature of gamma ray sources will, of course, require coordinated observations by X-ray telescopes in orbit and optical telescopes on Earth.

Returning to Earth, the final item on our agenda is the construction of permanent continent-sized radio astronomy arrays. The Australians plan a very long baseline array (the Australia Telescope), made up of the existing 64 m dish at Parkes, plus five 22 m dishes at Culgoora in New South Wales and another 22 m dish at Siding Spring. This will give a telescope with an effective diameter of 300 km. Eventually it will be linked by satellite to other radio telescopes at Carnarvon, Alice Springs, Hobart, Tidbinbilla and Fleurs (near Sydney) to give a 3000 km array.[14] The Canadian Long Baseline Array will consist of 8 32 m dishes arranged along the southern part of Canada, and a possible ninth radio telescope in the Yukon. It would operate at frequencies inbetween 611 MHz and 22GHz (wavelengths 49 cm to 1.3 cm) and cost about $70 million to build.

While for the Americans the construction of the Advanced X-ray Astrophysics Facility is ranked first by the Astronomy Survey Committee, and the National New Technology Telescope is ranked third, the second priority is the Very Long Baseline Array (VLBA), to consist of 10 25 m (82 foot) radio telescopes spread out over the United States from Westford, Massachusetts, to Hawaii and controlled from an operations center near the Very Large Array (VLA) in New Mexico. The VLBA will cost $68 million. In 1985 some $9 million was allocated for the start of construction. It is expected to be completed in the early 1990s. Given its operating frequency range of 325 MHz to 43 GHz (wavelengths inbetween 90 cm and 7 mm), it will allow a resolution as good as 200 *micro* arc seconds.[15] Should there be a huge black hole with an accretion disk in the center of our galaxy, the VLBA may allow us to make a map of it.

Each of these three radio astronomy arrays will be designed to be 'transparent' to the users so that they may go about their scientific business without having to be familiar with all the technicalities of the complex electronic and computer systems required. While this is easier said than done, much of the groundwork has been laid with projects such as the VLA. It is part of the cosmic scheme of things to allow the Canadian, Australian and American VLBAs to be linked together, giving a telescope almost as large as the Earth with the equivalent light-gathering power of a 146 m (480 foot) dish.

Without the construction of new state-of-the-art scientific equipment and the vigorous pursuit of answers to questions such as those outlined at the beginning of this chapter, astronomical science would only be left with the option to repeat experiments or review observations of the past. This does allow for any previous errors in analysis to be identified and corrected and for the models of cosmic objects to be revised at leisure. However, without the will or resources to take scientific inquiry to the next plateau we in fact regress, as did those ancient and medieval astronomers who relied on old observations instead of new ones made with more advanced equipment. May we fulfill our potential to expand the scope of our endeavors and successfully meet the challenges encountered in our quest to understand the universe which we are privileged to contemplate.

Appendix

The electromagnetic spectrum

There are many different kinds of electromagnetic radiation (light). Light waves, or photon packets (however you wish to think of them) are distinguished equivalently by the wavelength of the waves, the frequency of the waves, or the energy of the photons. The relationships described here are fundamental. First, the product of the wavelength of light and the frequency of the waves equals the speed of light:

$$\lambda v = c$$

If we use centimeter-gram-seconds (cgs) units, the wavelength λ is measured in cm, the frequency v is expressed in cycles per second, and the speed of light c is 3×10^{10} cm/s.

From simple quantum mechanics the energy of a photon is given by:

$$E = hv$$

The proportionality constant h is called Planck's constant and (in cgs units) is equal to 6.626×10^{-27} erg s. The photon energies would then be measured in ergs.

The first two relationships are equivalent to:

$$E = hc/\lambda$$

Astronomers usually express optical wavelengths not in cm, but in Ångströms, where $1 \text{ Å} = 10^{-8}$ cm. The optical slice of the electromagnetic spectrum is very small indeed, spanning approximately 3500–7000 Å. Infrared astronomers measure their photons by the wavelength in microns ($1 \mu m = 10^{-4}$ cm), or by wave numbers, which are the number of waves per unit distance in the direction of propagation. For example, if $\lambda = 2.122 \mu m$ (or 21220 Å), there are 4713 waves/cm.

Table A.1. *Regions of the electromagnetic spectrum*

↑	λ	ν(Hz)	E(eV)
radio waves			
$\updownarrow$	1 mm	3.0×10^{11}	1.24×10^{-3}
far infrared			
$\updownarrow$	20 μm	1.5×10^{13}	6.20×10^{-2}
near infrared			
$\updownarrow$	7000 Å	4.3×10^{14}	1.77
visible			
$\updownarrow$	3500 Å	8.6×10^{14}	3.54
ultraviolet			
$\updownarrow$	1000 Å	3.0×10^{15}	12.4
extreme ultraviolet			
$\updownarrow$	100 Å	3.0×10^{16}	124
X-rays			
$\updownarrow$	0.1 Å	3.0×10^{19}	1.24×10^5
gamma-rays			
↓			

Radio astronomers tend to think of their light waves in terms of frequency. The 21 cm wavelength photons of neutral hydrogen have a frequency of 1.429×10^9 cycles/s, or a frequency of 1429 megaHertz (MHz).

X-ray and gamma ray astronomers think of their photons in terms of energy, measured not in ergs, but in electron volts (1 eV = 1.6 $\times 10^{-12}$ ergs). The X-ray region is typically taken to be the wavelength regime from 100 Å to 0.1 Å, or, photon energies ranging from 1.24×10^2 eV to 1.24×10^5 eV. 1000 eV is a kilo-electron volt (1 KeV), 1 million electron volts is an MeV, and 10^9 eV is a GeV. If an electron and its antiparticle, the positron, meet and annihilate, they form two 0.511 MeV photons. Conversely, if high energy photons (> 1.022 MeV) are being produced in a cosmic source, electron–positron pairs can be produced from the interaction of the photons with atomic nuclei.

Table A.1 gives the approximate regimes of the electromagnetic spectrum. The boundaries are only for convention. For example, to be exact, visible photons are those that the average human eye can detect. Some people have eyes that can go slightly beyond the red or violet ends of the spectrum. Also, 'optical' astronomers can carry out ground-based observations on some tall mountains down to a wavelength of 3000 Å.

Spectra are classified into three basic types: (*a*) continuous spectra: (*b*) emission line spectra; and (*c*) absorption line spectra. Light emitted by a hot solid, liquid or dense gas forms a continuous, or smoothly varying spectrum. The wavelength at which such a 'black body' spectrum is brightest depends on the temperature of the object. For example, cooler stars are red, and hotter stars are blue. Light from a hot, rarefied gas, such as that in a switched-on mercury vapor lamp, gives off little continuous radiation, but instead an emission line (or bright line) spectrum whose lines correspond to the composition, density, and temperature of the gas. If we observe a source of black body radiation that is surrounded by cooler gas (or with cooler gas lying along the line of sight), we observe an absorption line spectrum whose dark lines (superimposed on the continuous spectrum) correspond to the composition, density and temperature of the cooler gas along the line of sight.

Clouds of molecules often give us molecular band spectra consisting of many closely spaced lines produced by the vibrations and rotations of the molecules. These lines need not be caused by the absorption of photons. Collisions between molecules can lead to line emission. Shock waves are a special case of this process.

Synchrotron emission is caused by the acceleration of charged particles in regions where there are magnetic fields. This is frequently observed by radio astronomers.

Notes

The following abbreviations have been used in these endnotes: *DSB* the *Dictionary of Scientific Biography*; *EB11*, *Encyclopaedia Britannica*, *11th ed., 1910–11*; *CDSB*, the *Concise Dictionary of Scientific Biography*; *GHA 4A*, *General History of Astronomy*, vol. 4, Part A (Ref. 6–21); Lang and Gingerich, *A Source Book in Astronomy and Astrophysics, 1900–1975* (Ref. 6–24); KS, Kellermann and Sheets (Ref. 8–93).

Preface

1. Quoted by Simon Newcomb, *The Reminiscences of an Astronomer*, Boston and New York: Houghton, Mifflin, 1903, p. 309.
2. Durant, Will and Durant, Ariel, *The Lessons of History*, New York: Simon & Schuster, 1968, p. 13.
3. See Mark Washburn, review of *The Right Stuff*, in *Sky and Telescope*, **67** (2), February 1984, pp. 136–7.

Chapter 1

1. See articles on 'Alexander the Great', 'Alexandria', 'Egypt', and 'Ptolemies', in *EB11*.
2. Sarton, George, *Introduction to the History of Science, Volume 1: From Homer to Omar Khayyám*, Huntington, New York: Robert E. Krieger, 1975 reprint, pp. 150, 158, 164, 172, 189. See also 'Libraries' in *EB11*, vol. 16, p. 546, and Parsons (Ref. 25 below) pp. 71, 349.
3. Grant, Robert, *History of Physical Astronomy*, London: Bohn, [1852], pp. 211–12.
4. Forbes, Eric G., *Greenwich Observatory. Volume 1: Origins and Early History (1675–1835)*, London: Taylor & Francis, 1975, pp. 7, 27.
5. Toomer, G. J., 'Ptolemy', in *DSB*, vol. 11, pp. 186–206, on p. 187. See also Sarton (Ref. 2), pp. 273, 545, 562.
6. Neugebauer, Otto, *The Exact Sciences in Antiquity*, Providence, Rhode Island: Brown University Press, 1957, quoting (David) Hilbert, p. 145.
7. The leader of the anti-Ptolemy contingent is Robert R. Newton, whose book *The Crime of Claudius Ptolemy* (Baltimore, 1977) resumed the debate of plagiarism in scientific matters. A debate on this is to be found in the References 8–10.
8. Gingerich, Owen, 'Was Ptolemy a Fraud?' *Quarterly Journal of the Royal Astronomical Society*, **21**, 1980, pp. 253–66.
9. Newton, Robert R., 'Comments on "Was Ptolemy a Fraud?" by Owen Gingerich,' ibid., pp. 388–99.
10. Gingerich, Owen, 'Ptolemy revisited: a reply to R. R. Newton', ibid., **22**, 1981, pp. 40–4.
11. A new edition of the *Almagest* has been edited by G. J. Toomer, *Ptolemy's Almagest*,

London: Duckworth & New York: Springer-Verlag, 1984, 693 p. This has been reviewed by Owen Gingerich in *Nature*, **308**, 1984, pp. 789–90.

12. Toomer, G. J., 'Hipparchus', in *DSB*, vol. 15 (Supplement), pp. 207–24, on p. 208.

13. According to C. W. Allen, *Astrophysical Quantities*, 3rd edn, London: Athlone, 1976, p. 141. The *longitude* of the 'Sun's apogee' in a geocentric system is the same as the longitude of the 'Earth's perihelion' (closest distance to the Sun) in a heliocentric system.

14. Neugebauer (Ref. 6), pp. 69, 185; Toomer (Ref. 12), pp. 216–17.

15. See Newton (Ref. 9), pp. 392–3; and Pannekoek, A., 'Ptolemy's precession', in *Vistas in Astronomy*, **1**, 1955, pp. 60–6. Dreyer (Ref. 20 below), pp. 346–8. performed the same calculation. The mean for the 18 stars according to him is 46″.6/year, with the lowest 7 values giving 37″.1 and the other 11 stars giving 52″.6.

16. Dreyer, J. L. E., *Tycho Brahe: A Picture of Scientific Life and Work in the Sixteenth Century*, Gloucester, Mass.: Peter Smith, 1977 reprint, pp. 265, 347.

17. Delambre, J. -B. J., *Histoire de l'Astronomie Ancienne*, 1817, vol. ii, p. 250 ff.

18. Newton (Ref. 7).

19. Boll, Franz, 'Die Sternkataloge des Hipparch und des Ptolemaios', *Biblioteca mathematica*, 3rd ser., **2**, 1901, pp. 185–95.

20. Dreyer, J. L. E., 'On the Origin of Ptolemy's Catalogue of Stars', *Monthly Notices of the Royal Astronomical Society*, **77**, 1917, pp. 528–39, and **78**, 1918, pp. 343–9.

21. See Toomer (Ref. 5), pp. 192–3; and Simon Newcomb's article on 'Moon' in *EB11*, vol. 18, p. 805.

22. Repsold, J. A., *Zur Geschichte der Astronomischen Messwerkzeuge von Purbach bis Reichenbach, 1450 bis 1830*, Leipzig: Wilhelm Engelmann, 1908, pp. 2–6; King, Henry C., *The History of the Telescope*, New York: Dover, 1979 reprint, pp. 6–9; and Lady Huggins' articles on 'Armilla' and 'Astrolabe' in *EB11*, vol. 2, pp. 575–6 and 795.

23. Sarton (Ref. 2), p. 348.

24. Gibbon, Edward, *The Decline and Fall of the Roman Empire*, New York: Modern Library, [1932?]. In vol. ii, p. 58 (Chapter XXVIII) Gibbon states: 'Theophilus proceeded to demolish the temple of Serapis . . . The valuable library of Alexandria [at the Serapeum] was pillaged or destroyed; and near twenty years afterwards, the appearance of the empty shelves excited the regret and indignation of every spectator whose mind was not totally darkened by religious prejudice'. The editor of this edition of Gibbon, Oliphant Smeaton, says in a note on this page that Gibbon was wrong, that 'only the sanctuary of the god . . . was levelled with the ground, and the library, the halls and the other buildings in the consecrated ground remained standing long afterwards . . . [W]e must conclude that the library in the Serapeum existed down to AD. 638'.

25. Parsons, Edward Alexander, *The Alexandrian Library: Glory of the Hellenic World*, Amsterdam: Elsevier, 1952, pp. 356–70.

26. During Caesar's Alexandrian war the 'mother' library was destroyed when flames from the harbor spread to the Brucheum quarter of the city. Antony presented Cleopatra with the library from Pergamum to make up for this loss. During the time of Aurelian in AD 273, the Brucheum quarter may have been destroyed, or at

least it declined in importance. From this time until at least 390 the literary center was in the Rhakotis section, at the Serapeum. See Smeaton's note on p. 57 of the edition of Gibbon (Ref. 24); See also 'Libraries' in *EB11*, vol. 16, p. 546.

27. Socrates Scholasticus, *Ecclesiastical History*, London: A. & J. Churchill, 1709, p. 376 (Book vii, Chapter 15).

28. Gibbon (Ref. 24), vol. ii, pp. 57–8 (Chapter XXVIII).

29. Ibid., p. 815 (Chapter XLVII).

30. 'Cyril' in *EB11*, vol. 7, p. 706.

31. Quoted by Kingsley, Charles, *Hypatia*, London: J. M. Dent, 1907, p. 429 (Chapter XXX). Kingsley's novel was first published in 1853.

32. McCutcheon, Robert A., *The purge of Soviet astronomy: 1936–37 with a discussion of its background and aftermath*, Georgetown University Master's thesis, 1985, 2 vols., 393 p. See abstract in *Bulletin of the American Astronomical Society*, **16** (2), 1984, p. 547.

33. For more on Saint Cyril see Gibbon (Ref. 24), vol. ii, pp. 814–30 (Chapter XLVII).

34. See 'Synesius' in *EB11*, vol. 26, p. 294; and Kingsley (Ref. 31) p. 291 (Chapter XXI).

35. Quoted by Hogarth, David George, 'Alexandria: History', *EB11*, vol. 1, p. 570.

36. Gibbon (Ref. 24) vol. iii, pp. 176–7 (Chapter LI).

Chapter 2

1. Sarton, George, *Introduction to the History of Science*, Huntington, New York: Robert E. Krieger, 1975 reprint, vol. 1, pp. 545–6, 557–8, 562–8.

2. Ibid., pp. 716–17.

3. Ibid., p. 603.

4. Ibid., p. 716.

5. Quoted by Rockhill, William Woodville, *The Journey of William of Rubruck to the Eastern Parts of the World, (1253–5), as Narrated by Himself, with Two Accounts of the Earlier Journey of John of Pian de Carpine*, London: Hakluyt Society, 1900, pp. xv–xvi. For detailed information on the Mongols see Vernadsky, George, *The Mongols and Russia*, New Haven, Connecticut: Yale University Press, 1953; Howorth, Henry H., *History of the Mongols from the 9th to the 19th Century*, London: Longman's, Green, & Co., 1880; Bretschneider, Emilii Vasilevich, *Medieval Researches from Eastern Asiatic Sources. Fragments towards the Knowledge of the Geography and History of Central and Western Asia from the 13th to the 17th Century*, New York: Barnes and Noble, 1967; and Douglas (Ref. 8, below).

6. Quoted in *The Horizon History of Russia*, New York: American Heritage, 1970, p. 72.

7. Ibid., p. 68.

8. Douglas, Robert Kennaway, 'Mongols,' in *EB11*, vol. 18, pp. 712–19.

9. See 'Maragha', in *EB11*, vol. 17, p. 667; and Sayili, Aydin, *The Observatory in Islam and its Place in the General History of the Observatory*, Ankara: Turkish Historical Society, 1960. Arno Press (New York) published a reprint of Sayili's book in 1981.

10. Sayili (Ref. 9), p. 199.

11. Nasr, Seyyed Hossein, 'Muḥammad ibn Muḥammad ibn al-Ḥasan al-Ṭūsi', in *DSB*,

vol. 13, pp. 508–14. See Saliba, George, 'The first non-Ptolemaic astronomy at the Maraghah school', *Isis*, **70**, 1979, pp. 571–6. Also, Kennedy, E. S., 'Late medieval planetary theory', *Isis*, **57**, 1966, pp. 365–78.

12. Sarton (Ref. 1), vol. 3, part 2, pp. 1467–74; see also Goldsmid, Frederick John, 'Timur', in *EB11*, vol. 26, pp. 994–5.

13. Sarton (Ref. 1), vol. 3, part 2, pp. 1120, 1469.

14. Kari-Niazov, T. N., 'Ulugh Beg', in *DSB*, vol. 13, pp. 535–7.

15. Baily, Francis, 'The Catalogues of Ptolemy, Ulugh Beigh, Tycho Brahe, Halley, Hevelius, deduced from the best authorities, etc.', *Memoirs of the Royal Astronomical Society*, **13**, 1843, pp. 19–28, (79)–(126).

16. Sheglov, V. P., *Jan Hevelius: the star atlas*, Tashkent: Fan Press, 1968, pp. XI–XXIX.

17. Sayili, Aydin, *Uluğ Bey Ve Semerkanddeki İlim Faaliyeti Hakkinda Giyasüddin-i Kâsi'nin Mektubu* (Ghiyâth al Dîn al Kâshî's 'Letter on Ulugh Bey and the Scientific Activity in Samarqand'), Ankara: Türk Tarih Kurumu Basimevi, 1960, pp. 32–53, 93–110, on p. 104.

18. Ibid., p. 98.

19. Ibid., pp. 94–5.

20. Ibid., p. 95.

21. Ibid., p. 104.

22. Ibid., p. 106.

23. Vyatkin, V. L., 'Ochet o raskopkakh observatorii Mirza Ulugbeka v 1908 i 1909 godakh', *Izvestiia Russkago komiteta dlia izucheniia srednei i vostochnoi azii*, 2nd. ser. (1912), no. 11.

24. Shishkin, V. A., *Trudy Instituta Istorii i arkheologii*, Acad. Scis. Uzbek S. S. R., vol. 5, Tashkent, 1953.

25. Kari-Niazov, T. N., *Astronomicheskaia shkola Ulugbeka*, 2nd ed., Tashkent, 1967.

26. Kari-Niazov (Ref. 14), p. 536; Sheglov (Ref. 16), pp. XV, XIX–XX.

27. Kari-Niazov (Ref. 14), p. 536; Sayili (Ref. 17), pp. 106–7.

28. Kari-Niazov (Ref. 14), p. 537.

29. See Sheglov (Ref. 16), pp. XV–XVI, XX–XXI; and Baily (Ref. 15), pp. 25–6.

30. Sarton (Ref. 1), vol. 1, pp. 665–6

31. Sayili (Ref. 17), p. 103.

32. Ibid., pp. 109–10.

33. See Howorth (Ref. 5), vol. 2, pt. 2, pp. 687–8; Bretschneider (Ref. 5), vol. 2, pp. 262–3; Baily (Ref. 15), pp. 23–4; and Sheglov (Ref. 16), p. XI.

34. Kari-Niazov (Ref. 14), p. 537.

35. See Repsold, J. A., *Zur Geschichte der Astronomischen Messwerkzeuge von Purbach bis Reichenbach, 1450 bis 1830*, Leipzig: Wilhelm Engelmann, 1908, pp. 6–11; Sédillot, L., *Mémoire sur les instruments astronomiques des Arabes*, Paris, 1841.

36. Sarton (Ref. 1), vol. 1, p. 667; Sheglov (Ref. 16), p. XIX.

37. Maddison, Francis R., 'A 15th century Islamic spherical astrolabe', *Physis. Rivista di storia della scienza*, **IV**, Fasc. 2, 1962, pp. 101–9.

38. North, John D., 'The astrolabe', *Scientific American*, **270**, January 1974, pp. 96–106; and Webster, Roderick S., 'The astrolabe: some notes on its history, construction, and use', Box 157, Lake Bluff, Illinois 60044: Paul MacAlister, 1974.

The latter comes with an astrolabe kit available from the publisher for $20.

39. Sarton (Ref. 1), vol. 1, p. 193.

40. Arabic translation of *Planisphaerium* by Maslama al-Majrītī (d. 1007). Latin translation of this by Hermann of Carinthia (1143). See 'Ptolemy' in *DSB*, vol. 11, pp. 197–8. For German translation see Drecker, J. D., 'Das Planisphaerium des Claudius Ptolemaeus,' *Isis*, **9**, 1927, pp. 255–78.

41. Drecker, J. D., 'Des Philoponus Schrift über das Astrolab', *Isis*, **11**, 1928, pp. 15–44.

42. Neugebauer, O., 'The early history of the astrolabe', *Isis*, **40**, no. 3, 1949, pp. 240–56. Reprinted in Neugebauer, O., *Astronomy and History: Selected Essays*, New York: Springer, 1983, pp. 278–94.

43. Carter, Tom, 'Geoffrey Chaucer: Amateur Astronomer?' *Sky and Telescope*, **63** (3), March 1982, pp. 246–7.

44. One should beware of spending a lot of money on a 'genuine' antique astrolabe, for many of them are bogus. See Gingerich, Owen, 'Fake astrolabes', *Sky and Telescope*, **63** (5), May 1982, pp. 465–8.

Chapter 3

1. Sarton, George, in his *Introduction to the History of Science. Volume I: From Homer to Omar Khayyám*, Huntington, New York: Robert E. Krieger, 1975 reprint, praises the intellectual activity at Cordova and attributes it to the patronage of its rules. 'Arabic Spain became under ['Abd al-Rahmān III, 8th Umayyad caliph of Cordova from 912 to 961] and his successors one of the most civilized and best governed countries of the time' (p. 628). Al-Hakam II, 9th Umayyad caliph of Cordova from 961 to 976, is said to have had a library containing 400 000 volumes (p. 658). By the second half of the eleventh century, 'outside of Persia, the only center of intellectual progress in Islam was Spain, but the heyday of Cordova was already over' (p. 746). However, the development of astronomy by al-Zarqālī, who flourished in Toledo, is to be considered a definite step forward (pp. 746–7, 758–9).

2. Sarton, George, *Six Wings: Men of Science in the Renaissance*, Bloomington: Indiana University Press, 1957, p. 90.

3. Article on Gerard of Cremona in CDSB, 281–2.

4. Sarton (Ref. 2), p. 257.

5. Forbes, Eric G., *Greenwich Observatory: ... Origins and early history (1675–1835)*, London: Taylor & Francis, 1975, p. 1.

6. Pedersen, Olaf, 'Some early observatories', *Vistas in Astronomy*, **20**, 1976, pp. 17–28.

7. Sarton (Ref. 2), pp. 63, 259.

8. Ibid., p. 257.

9. Repsold, Johann A., *Zur Geschichte der Astronomischen Messwerkzeuge von Purbach bis Reichenbach (1450 bis 1830)*, Leipzig: Wilhelm Engelmann, 1908, p. 16; Pedersen (Ref. 6), p. 18.

10. *Tycho Brahe's Description of his instruments and scientific work as given in Astronomiae Instauratae Mechanica (Wandesburgi 1598)*, translated and edited by Hans Raeder,

Elis Strömgren, and Bengt Strömgren, Copenhagen, 1946, p. 131; Pedersen (Ref. 6), p. 18.

11. Repsold (Ref. 9), p. 18; Pedersen (Ref. 6), p. 27 (note 37).
12. Repsold (Ref. 9), p. 21 (original in German).
13. Ibid.
14. Pedersen (Ref. 6), p. 19, states that Tycho's combined annual income of 2400 *daler* was about 1% of the government's annual income. Thus, 75000/(2400 × 100) is about 30%.
15. Denmark's national budget for 1980 showed revenues of $15.4 billion, according to the 1984 *World Almanac and Book of Facts*, p. 491.
16. Quoted by Alexander von Humboldt, *Cosmos: a sketch of a physical description of the universe*, New York: Harper, 1860, vol. 3, p. 152.
17. Dreyer, J. L. E., *Tycho Brahe: A Picture of Scientific Life and Work in the Sixteenth Century*, Gloucester, Mass.: Peter Smith, 1977 reprint, pp. 108–13; see also Pedersen (Ref. 6), p. 19. Dreyer's book, originally published in Edinburgh in 1890, is the standard definitive biography of Tycho (at least until Victor Thoren's new work is published).
18. Pedersen (Ref. 6), p. 21.
19. See Raeder *et al.* (Ref. 10), and Dreyer (Ref. 17), p. 315 ff. Many of Tycho's instruments are described and pictured in Repsold (Ref. 9), pp. 21–30.
20. Raeder *et al.* (Ref. 10), p. 9; Repsold (Ref. 9), p. 22. The former adopt 1 cubit = 388 mm, while the latter adopts 1 cubit = 417 mm.
21. Raeder *et al.* (Ref. 10), p. 126.
22. Ibid., pp. 128–32.
23. Pedersen (Ref. 6), p. 19.
24. Hellman, C. Doris, 'Tycho Brahe', in *DSB*, vol. 2, pp. 401–16, on p. 406. See also Dreyer (Ref. 17), pp. 117–27, 381–4.
25. Sarton (Ref. 2), p. 64.
26. Gingerich, Owen, 'Tycho Brahe and the Great Comet of 1577', *Sky and Telescope*, **54** (6), December 1977, pp. 452–8.
27. Sarton (Ref. 2), pp. 64–5; Hellman (Ref. 24), pp. 407–8.
28. Dreyer (Ref. 17), pp. 191–2.
29. See also Baily, Francis, 'The Catalogues of Ptolemy, Ulugh Beigh, Tycho Brahe, Halley, Hevelius, deduced from the best authorities, etc.', *Memoirs of the Royal Astronomical Society*, **13**, 1843, pp. 29–34, (127)–(166).
30. Sarton (Ref. 2), pp. 65, 259.
31. See Dreyer (Ref. 17),p. 335. Tycho derived two different refraction tables, one for the Sun, the other for the stars. For solar observations he believed refraction negligible above elevations of 45°, while for stars the critical value was an elevation angle of 20°. His refraction correction errors for solar observations were partly cancelled out by his adopted value (incorrectly large) of the solar parallax.
32. Dreyer (Ref. 17), p. 191.
33. See Mayer, Ben, 'Touring the Jai Singh Observatories', *Sky and Telescope*, **58** (1), July 1979, pp. 6–10. Jai Singh (1686–1743), an Indian ruler and patron of science, constructed a number of large, naked eye instruments from 1724 to 1737.

34. See Forbes (Ref. 5), pp. 1, 6, and Kopal, Zdenek, 'Ole Christensen Römer', in *DSB*, vol. 11, pp. 525–7. Dreyer (Ref. 17), pp. 358–60, 388–9, claims that there is no substantiation of the claim for a meridian error of 15′.
35. Raeder *et al.* (Ref. 10), pp. 29–30.
36. Dreyer (Ref. 17), pp. 387–8. Tycho's coordinates are compared to James Bradley's observations (reduced by Auwers).
37. Sarton (Ref. 1), p. 666, and Simon Newcomb's article on the 'Moon' in the *EB11*, dispel the notion that Abū-l-Wafā (*c.* 940 to 997 or 998) preceded Tycho Brahe in the discovery of the variation.
38. Thoren, Victor, E., 'Tycho and Kepler on the lunar theory', *Publications of the Astronomical Society of the Pacific*, **79**, October 1967, pp. 482–9.
39. Dreyer (Ref. 17), pp. 26–7. See also Ashbrook, Joseph, 'Tycho Brahe's Nose', *Sky and Telescope*, **29**(6), June 1965, pp. 353, 358, where we also learn of Dr Heinrich Matiegka's 'Report on the investigation of the skeleton of Tycho Brahe', presented to the Royal Bohemian Academy of Sciences on 11 October 1901 and published in the academy's publications. The name of the other Danish nobleman is given as 'Manderupius Pasbergius' by August De Morgan, 'Assorted paradoxes', in *The World of Mathematics*, James R. Newman, ed., New York: Simon and Schuster, 1956, pp. 2369–82, on p. 2375.
40. Repsold (Ref. 9), p. 29.
41. Hanson, Norwood Russell, *Patterns of Discovery*, Cambridge University Press, 1972, pp. 73–85.
42. Hall, A. Rupert, *The Scientific Revolution, 1500–1800: the formation of the modern scientific attitude*, Boston: Beacon Press, 1966, p. 121.
43. Ibid., quoted on p. 247.
44. Ibid., pp. 258–76.
45. Quoted by Giorgio de Santillana, *The Crime of Galileo*, New York: Time, 1962, p. 36.
46. Shapere, Dudley, 'Meaning and scientific change', in *Scientific Revolutions*, ed. Ian Hacking, Oxford University Press, 1981, pp. 28–59, on p. 33.

Chapter 4

1. Storr, Francis, 'Academies', in *EB11*, vol. 1, pp. 97–105.
2. Cohen, I. Bernard, *Album of science: from Leonardo to Lavoisier. 1450–1800*, New York: Scribner's, 1980, p. 234 ff.
3. Wolf, C[harles Joseph Etienne], *Histoire de l'Observatoire de Paris de sa Foundation à 1793*, Paris: Gauthier-Villars, 1902, p. 116.
4. Barthalot, Raymonde, 'The story of Paris Observatory', *Sky and Telescope*, **59** (2), February 1980, pp. 100–7.
5. Grant, Robert, *History of physical astronomy*, London: Bohn, [1852], pp. 457–8.
6. Winterhalter, Albert G., *The International Astrophotographic Congress and a Visit to Certain European Observatories and other Institutions, Appendix I to Washington Observations for 1885*, Washington, DC: Government Printing Office, 1889, p. 96.
7. Boyer, Carl B., 'Early estimates of the velocity of light', *Isis*, **33**, 1941, pp. 24–40. See also: Cohen, I. Bernard, 'Roemer and the first determination of the velocity of light (1676)', *Isis*, **31**, 1939, pp. 327–79.

8. Forbes, Eric G., *Greenwich Observatory: ... Origins and Early History (1675–1835)*, London: Taylor & Francis, 1975, p. 91 ff. For complete citation see Ref. 15, below.

9. King, Henry C., *The History of the Telescope*, New York: Dover, 1979 reprint, p. 48 ff.

10. Ibid., and notes by S. Grillot and J. Levy for the *General History of Astronomy*.

11. Taton, Rene, series of articles on the Cassinis, in *DSB*, vol. 3, pp. 100–8. See also: Forbes (Ref. 8), pp. 143–7, Grant (Ref. 5), pp. 147–9; Barthalot (Ref. 4), p. 102; and Warner, Brian, *Astronomers at the Royal Observatory, Cape of Good Hope*, Capetown & Rotterdam: Balkema, 1979, pp. 50–3, 56–7.

12. Levy, J., 'La creation de la Connaissance des Temps', *Vistas in Astronomy*, **20**, 1976, pp. 75–7. See also Forbes (Ref. 8), pp. 122–5; Hall, A. Rupert, *The Scientific Revolution, 1500–1800: The Formation of the Modern Scientific Attitude*, Boston: Beacon Press, 2nd edn, 1962, pp. 203–4.

13. Spencer Jones, Harold, 'John Couch Adams and the discovery of Neptune', in *The World of Mathematics*, James R. Newman, ed., New York: Simon & Schuster, 1956, pp. 822–39. See also Grant (Ref. 5), pp. 164–201.

14. See Winterhalter (Ref. 6), in particular pp. 48–9. See also: Van Biesbroeck, G., 'Star catalogues and charts', in *Basic astronomical datas* ed. K. Aa. Strand, Chicago & London: University of Chicago Press, 1963, pp. 471–80; Ingrao, Hector C. & Kasparian, Elaine, 'Photographic star atlases', in *Sky and Telescope*, **34**, (5), November 1967, pp. 284–7; and *Transactions of the International Astronomical Union*, XIVB, Dordrecht: Reidel, 1971, pp. 170–8.

15. *Greenwich Observatory. One of Three Volumes by Different Authors Telling the Story of Britain's Oldest Scientific Institution. The Royal Observatory at Greenwich and Herstmonceux. 1675–1975. Volume 1: Origins and Early History (1675–1835)*, by Eric G. Forbes, xv + 204 p. *Volume 2: Recent History (1836–1975)*, by A. J. Meadows, xi + 135 p. *Volume 3: Buildings and Instruments*, by Derek Howse, xix + 178 p., London: Taylor & Francis, 1975.

16. See Meadows (Ref. 15), p. 4.

17. Private communication from Janet Dudley, RGO librarian and archivist, 10 August 1982.

18. McCrea, W. H., 'The Royal Greenwich Observatory, 1675–1975: Some of its external relations', *Quarterly Journal of the Royal Astronomical Society*, **17**, 1976, pp. 4–24, on p. 5.

19. Forbes (Ref. 15), pp. 13–14.

20. Sadler, D. H., 'Lunar distances and the Nautical Almanac', *Vistas in Astronomy*, **20**, 1976, pp. 113–21.

21. Forbes, Eric G., 'The origins of the Greenwich Observatory', *Vistas in Astronomy*, **20**, 1976, pp. 39–50. See also Forbes (Ref. 15), pp. 17–19, and Grant (Ref. 5), pp. 459–60.

22. Quoted in Forbes (Ref. 15), p. 19.

23. Ibid., p. 22.

24. Quoted by Howse, Derek, 'Restoration at Greenwich Observatory,' *Sky and Telescope*, **40** (1), July 1970, pp. 4–9, on p. 5.

25. The primary source of information on the early history of Greenwich Observatory is Francis Baily's *An account of the Revd John Flamsteed, the first Astronomer–Royal;*

etc. To which is added his British Catalogue of Stars, Corrected and Enlarged, London, 1835. This has been reprinted by Dawsons of Pall Mall, London, 1967, but without Flamsteed's catalogue. See also *The Preface to John Flamsteed's Historia Coelestis Britannica or British Catalogue of the Heavens*, ed. Allan Chapman, London: National Maritime Museum, 1982, vi + 222 p.

26. Chapman, Allan, 'Astronomia practica: The principal instruments and their uses at the Royal Observatory', *Vistas in Astronomy*, **20**, 1976, pp. 141–56. See also Grant (Ref. 5), pp. 467–8, Forbes (Ref. 15), pp. 29–32, and Howse (Ref. 15), pp. 75–9.

27. Laurie, P. S., 'The observer's life', *Vistas in Astronomy*, **20**, 1976, pp. 187–90. See also Forbes (Ref. 15), pp. 155–6.

28. See Forbes (Ref. 15), pp. 49–50.

29. Quoted in a book review of Baily (Ref. 25), by Owen Gingerich, *Sky and Telescope*, **34**, (1), July 1967, pp. 35–7.

30. Grant (Ref. 5), p. 476.

31. Howse (Ref. 15), pp. 21–4.

32. Quoted by Forbes (Ref. 15), p. 82.

33. Grant (Ref. 5), p. 479.

34. Brown, Lloyd A., 'The longitude', in Newman (Ref. 13), pp. 780–819.

35. See Forbes (Ref. 15), p. 91 ff. and Grant (Ref. 5), pp. 336–42.

36. Smart, W. M., *Text-book on Spherical Astronomy*, Cambridge University Press, 5th edn, 1965, pp. 181–2, 219–20.

37. Ibid., pp. 226–34, 247.

38. Grant (Ref. 5), p. 484.

39. Chapman (Ref. 26), p. 141.

40. Meadows (Ref. 15), pp. 105–7.

41. Quoted by Forbes (Ref. 15), p. 171.

42. Clerke, Agnes, *A Popular History of Astronomy During the Nineteenth Century*, London: Adam & Charles Black, 4th edn, 1902, p. 4.

43. Quoted by Owen Gingerich, 'Herschel's 1784 autobiography', *Sky and Telescope*, **68** (4), October 1984, 317–319, on p. 318.

44. Hoskin, Michael, 'William Herschel', in *DSB*, vol. 6, pp. 328–36 and references therein.

45. Herrmann, Dieter B., *The History of Astronomy from Herschel to Hertzsprung*, Cambridge University Press, 1984. Contains much discussion of William Herschel's work.

46. Meadows (Ref. 15), p. 15.

47. Forbes (Ref. 15), p. 131.

48. Meadows (Ref. 15), p. 40.

49. Herrmann (Ref. 45), pp. 38–41, and Meadows (Ref. 15), pp. 44–6, 50–1.

50. Meadows (Ref. 15), pp. 48–50.

Chapter 5

1. Lincoln, W. Bruce, *The Romanovs: Autocrats of all the Russias*, New York: Dial Press, 1981, p. 239 ff.

2. *The Horizon History of Russia*, New York: American Heritage, 1970, p. 178.

3. Ibid., p. 171.

4. Grant, Robert, *History of physical astronomy*, London: Bohn, [1852], pp. 503–5.
5. Ashbrook, Joseph, 'The crucial years of Wilhelm Struve', *Sky and Telescope*, **25** (6), June 1963, pp. 326–7. See also Sokolovskaya, Z. K., 'Friedrich Georg Wilhelm Struve', in *DSB*, vol. 13, pp. 108–13.
6. Russia did not convert to the Gregorian calendar until 1918, by which time they lagged 13 days behind the rest of the western world. In the nineteenth century this lag amounted to 12 days. The Julian calendar is referred to as the Old System (OS); the Gregorian calendar is the New System (NS).
7. Abbe, Cleveland, 'Dorpat and Poulkova', *Annual Report of the Board of Regents of the Smithsonian Institution ... for the Year 1867*, Washington: Government Printing Office, 1872, pp. 370–90, on pp. 373–4.
8. Quoted by Novokshanova (Sokolovskaia), Z[inaida] K[uzminichna], *Vasilii IAkovlevich Struve*, Moscow: Izdatel'stvo 'Nauka', 1964, p. 56.
9. Struve, F[riedrich] G[eorg] W[ilhelm], *Description de l'Observatoire Astronomique central de Poulkova*, St Petersburg: Imperial Academy of Sciences, 1845, pp. 30–4 and 115—211.
10. Novokshanova (Ref. 8), pp. 66–78 and 231–2.
11. Abbe (Ref. 7), pp. 385–7.
12. Struve, 1845 (Ref. 9), p. 56.
13. Ibid., p. 53.
14. This work is quite rare. One set of the two volumes sold in 1982 for $1850. See Alan H. Batten, 'A recent valuable acquisition by the library of the Dominion Astrophysical Observatory', *Journal of the Royal Astronomical Society of Canada*, **76** (6), December 1982, pp. 382–91.
15. Struve, Otto Wilhelm, *Übersicht der Thätigkeit der Nicolai-Hauptsternwarte während der ersten 25 jahre ihres Bestehens*, St Petersburg: Imperial Academy of Sciences, 1865, 119 p. This work also appeared in a Russian translation.
16. Struve, Otto Wilhelm, *K 50-letiiu Nikolaevskoi Glavnoi astronomicheskoi observatorii*, St. Petersburg, 1889. This work also appeared in a German translation.
17. Polianskaia, L., *K stoletiiu Pulkovskoi observatorii, Krasnyi arkhiv*, vol. 4(95), 1939.
18. Novokshanova (Ref. 8), pp. 48–103.
19. Dadaev, Aleksandr Nikolaevich, *Pulkovo Observatory: an Essay on its History and Scientific Activity*, translated by Kevin Krisciunas, Springfield, Virginia: National Technical Information Service, NASA TM-75083, 1978, 239 p.
20. Krisciunas, Kevin, 'A short history of Pulkovo Observatory'. *Vistas in Astronomy*, **22** (pt. 1), 1978, pp. 27–37.
21. KHotinskii, M., 'Obzor sochineniia akademika V. IA. Struve', *Opisanie Glavnoi astronomicheskoi observatorii*, St Petersburg, 1847, p. 30. Quoted by Novokshanova (Ref. 8), p. 70. Original in Russian.
22. Dadaev (Ref. 19), pp. 15–16.
23. Struve, (F. G. W.), 'Notice sur l'instrument des passages de Repsold, etabli a l'Observatoire de Poulkova dans le premier vertical, et sur les resultats que cet instrument a donné pour l'évaluation de la constante de l'aberration,' reprinted from the *Bulletin scientifique publie par l'Academie imp. des Sciences de St-Petersbourg*, **10**, (14, 15 & 16), 1843, 21 p. + 2 plates.
24. Dadaev (Ref. 19), pp. 13, 126.

25. For Otto Wilhelm Struve's work on precession see *Monthly Notices of the Royal Astronomical Society*, **10**, 1850, pp. 100–9. Otto Wilhelm Struve's son Hermann also won the Gold Medal of the RAS, in 1903. The second Otto Struve (1897–1963), great-grandson of F. G. W. Struve, won it in 1944. See Alan H. Batten, 'The Struves of Pulkovo – a family of astronomers', *Journal of the Royal Astronomical Society of Canada*, **71** (5), October 1977, pp. 345–72.

26. Attributed to B. A. Gould by Simon Newcomb, *Reminiscences of an Astronomer*, Boston & New York: Houghton, Mifflin, 1903, p. 309.

27. See Moffett, Alan T., 'Argelander and the BD', *Sky and Telescope*, **29** (5), May 1965, pp. 276–8; Ashbrook, Joseph, 'How the BD was made', *Sky and Telescope*, **59** (4), April 1980, pp. 300–2; a general discussion of nineteenth century positional astronomy, including the *Bonner Durchmusterung*, is to be found in Herrmann, Dieter B., *The History of Astronomy from Herschel to Hertzsprung*, Cambridge University Press, 1984, pp. 20–53.

28. Bessel had predicted in 1844, on the basis of the non-uniform motion of the star Procyon, that it had an 'invisible' companion. In 1873 O. W. Struve claimed to have spotted a faint companion of the bright star. These observations were made with the Pulkovo 15 inch. However, the 26 inch refractor at the US Naval Observatory could show no such object. It turned out that Struve was mistaken in his identification of the faint star, for the first bona fide sighting was by Schaeberle with the Lick Observatory 36 inch in 1896. See Agnes Clerke, *A Popular History of Astronomy During the Nineteenth Century*, London: Adam and Charles Black, 4th edn, 1902, pp. 41–2, and Joseph Ashbrook, 'Companion of Procyon: some preliminary reports', *Sky and Telescope*, **28** (2), August 1964, p. 72.

29. Krisciunas, 1978 (Ref. 20), p. 34; A. A. Mikhaïlov, personal communication, 30 January 1982; *Astronomy and Astrophysics Abstracts*, **34**, (1983, Part 2), p. 74; *Soviet Astronomy*, **28** (2), March–April 1984, pp. 243–4; *Astronomy and Astrophysics Abstracts*, **37**, (1984, Part 1), p. 76; V. K. Abalakin, personal communication, 5 September 1987.

30. Clerke (Ref. 28), pp. 429–30.

31. Letter of O. W. Struve to Simon Newcomb (30 March 1879). The original is in German.

32. Letter of O. W. Struve to Mrs Newcomb (29 June 1885). The original is in English.

33. O. W. Struve, 1889 (Ref. 16), p. 5. Quoted by Dadaev (Ref. 19), p. 23.

34. Krisciunas, Kevin 'The end of Pulkovo Observatory's reign as the "astronomical capital of the world"' *Quarterly Journal of the Royal Astronomical Society*, **25** (3), September 1984, pp. 301–305.

35. Quoted in Kulikovsky, P. G., 'Fëdor Aleksandrovich Bredikhin', in *DSB*, vol. 2, pp. 432–5, on p. 434.

36. Letter of O. W. Struve to Simon Newcomb (16 April 1895). The original is in German.

37. See Dadaev (Ref. 19), and Nicolaidis, E., *Le Developpement de l'Astronomie en URSS 1917–1935*, Paris: Observatoire de Paris, 1984, 255 p.

Chapter 6

1. Wright, Helen, *Explorer of the Universe: a Biography of George Ellery Hale*, New York: E. P. Dutton, 1966, pp. 115–16.
2. Osterbrock, Donald E., *James E. Keeler: Pioneer American Astrophysicist and the Early Development of American Astrophysics*, Cambridge University Press, 1984, pp. 204–11.
3. Herrmann, Dieter B., *The History of Astronomy from Herschel to Hertzsprung*, Cambridge University Press, 1984, pp. 69 ff.
4. Osterbrock (Ref. 2), p. 2.
5. Jones, Bessie Zaban, and Boyd, Lyle Gifford, *The Harvard College Observatory. The First Four Directorships, 1839–1919*, Cambridge, Massachusetts: Belknap Press, 1971, quoted on p. 36. This is the standard work on the first 80 years of HCO.
6. Dick, Steven J., 'How the U.S. Naval Observatory began, 1830–65', *Sky and Telescope*, **60** (6), December 1980, pp. 466–71.
7. Rhynsburger, R. W., 'A historic refractor's 100th anniversary', *Sky and Telescope*, **46** (4), October 1973, pp. 208–14.
8. See Jones and Boyd (Ref. 5), pp. 37–8; Dreyer, J. L. E., 'Observatory', in *EB11*, vol. 19, pp. 953–61; and Zinszer, Harvey A., 'Famous early American observatories', *Transactions of the Kansas Academy of Science*, **47**, 1944, pp. 15–25.
9. Jones and Boyd (Ref. 5), pp. 48–51.
10. Ibid., quoted on p. 51.
11. Ibid., p. 54.
12. See articles by Owen Gingerich on George Phillips Bond and William Cranch Bond in *DSB*, vol. 2, pp. 284–5.
13. Jones and Boyd (Ref. 5), p. 67.
14. Ibid., p. 133.
15. See articles on 'Daguerre' and 'Photography' in *EB11*. Michael Broyles, whose pictures are to be found in Chapters 8 and 9 of this book, tells me that the wet-collodion plates have *never* been surpassed for the crispness of the images. They are no longer used because of their low sensitivity and the difficulty of handling them.
16. G. P. Bond to William Mitchell, 6 July 1857. Quoted by Jones and Boyd (Ref. 5), pp. 84–5.
17. Ibid., pp. 71–87; Hoffleit, Dorrit, 'The first star photograph', *Sky and Telescope*, **9** (9), July 1950, pp. 207–10; Herrmann (Ref. 3), pp. 81–7; *EB11* articles on 'Huggins, Sir William' and 'Photography'.
18. Jones and Boyd (Ref. 5), pp. 118–19.
19. Ibid., pp. 87, 148, 157.
20. Clerke, Agnes M., *A Popular History of Astronomy during the Nineteenth Century*, London: Adam & Charles Black, 4th edn, 1902, p. 171.
21. See Jones and Boyd (Ref. 5), pp. 176–444; Plotkin, Howard, 'Edward Charles Pickering', in *DSB*, vol. 10, pp. 599–601; Plotkin, Howard, 'Harvard College Observatory', in *General History of Astronomy. Volume 4: Astrophysics and Twentieth-Century Astronomy to 1950: Part A*, ed. Owen Gingerich, Cambridge University Press, 1984, pp. 122–4.
22. Gingerich, Owen, 'The first photograph of a nebula', *Sky and Telescope*, **60** (5), November 1980, pp. 364–6.

23. Quoted by Jones and Boyd (Ref. 5), p. 248.

24. Jones and Boyd (Ref. 5), pp. 236–40; Hertzsprung, Ejnar, 'On the radiation of stars,' translated by Harlow Shapley and Vincent Icke, in *A Source Book in Astronomy and Astrophysics, 1900–1975* ed. Kenneth R. Lang and Owen Gingerich, Cambridge, Mass.: Harvard University Press, 1979, pp. 208–11.

25. Under the direction of John S. Paraskevopoulos the Boyden station began operations in September 1927 at Mazelspoort, 14 miles ENE of Bloemfontein, South Africa. The final instrumentation was installed in 1933, which included a 60 inch Newtonian reflector, a 24 inch astrograph, a 13 inch refractor, a 10 inch photographic refractor, an 8 inch photographic refractor and some smaller instruments. The 24 inch astrograph was replaced in 1950 by a 32 inch Baker–Schmidt camera. Harvard ran the Boyden station through 1954. Beginning the following year the administration was shared with five European observatories: Armagh (Northern Ireland), Dunsink (near Dublin), Hamburg, Stockholm and Uccle (Belgium). There have also been two German observatories on the same site: the Bamberg Southern Station for variable star observations, equipped with a 16 inch photoelectric Cassegrain reflector; and the Heidelberg Station, with a 20-inch Cassegrain reflector. See *Scientiae* (printed in Pretoria, S. Africa) **9** (10), October 1968, p. 16; and Stoy, R. H., 'Astronomy', in *Standard Encyclopaedia of Southern Africa*, Cape Town: Nasou Ltd., 1970, vol. 1, pp. 583–9, on p. 588.

26. Jones and Boyd (Ref. 5), pp. 275–6; John Lankford, in his article 'Astronomical photography', *GHA 4A*, pp. 16–39, on p. 32, states:

    By a curious turn of events...the growth of astrophysics in the United States may have been stimulated as a consequence of non-participation in the *Carte du Ciel* while, at least to a degree, astrophysical research in Europe may have been retarded because the *Carte* absorbed funds and engrossed staff time that otherwise might have been allocated to astrophysics.

27. Leavitt, Henrietta S., 'Periods of twenty-five variable stars in the Small Magellanic Cloud', in Lang and Gingerich (Ref. 24), pp. 398–400.

28. Bok, Bart J., 'Harlow Shapley – cosmographer and humanitarian', *Sky and Telescope*, **44** (6), December 1972, pp. 354–7.

29. See 'The scale of the universe. Part 1: Harlow Shapley. Part 2: Heber D. Curtis', in Lang and Gingerich (Ref. 24), pp. 523–41; Hoskin, M. A., 'The "Great Debate": what really happened', *Journal for the History of Astronomy*, 7, 1976, pp. 169–82; Seeley, Daniel, and Berendzen, Richard, 'Astronomy's Great Debate', *Mercury*, **7** (4), July–August 1978, pp. 67–71, 88; Smith, Robert W., 'The Great Debate revisited', *Sky and Telescope*, **65** (1), January 1983, pp. 28–9.

30. Shapley's final annual report as director, quoted by Bok (Ref. 28) p. 356.

31. Struve, Otto, and Zebergs, Velta, *Astronomy of the 20th Century*, New York: Macmillan, 1962, p. 220.

32. Whitney, Charles A., 'Cecilia Payne-Gaposchkin: an astronomer's astronomer', *Sky and Telescope*, **59** (3), March 1980, pp. 212–14; Haramundanis, Katherine, ed., *Cecilia Payne-Gaposchkin: an Autobiography and Other Recollections*, Cambridge University Press, 1984. See also Dorrit Hoffleit's review of this book in *Sky and Telescope*, **68** (3), September 1984, pp. 225–7.

33. Meadows, A. J., *Greenwich Observatory . . . Vol. 2: Recent History (1836–1975)*, London: Taylor & Francis, 1975, p. 14.

34. Lick, Rosemary, *The Generous Miser: the Story of James Lick of California*, Menlo Park, California: Ward Ritchie Press, 1967, pp. xi–xvi (by C. D. Shane), 61–95. Be on the lookout for *Lick Observatory: the first century*, by Donald E. Osterbrock, John Gustafson and William Unruh [1988(?)].

35. Osterbrock (Ref. 2), pp. 37–8.

36. Quoted by R. Lick (Ref. 34), p. 90.

37. Osterbrock (Ref. 2), p. 38.

38. R. Lick (Ref. 34), p. 66.

39. Ibid., pp. 90–3.

40. Newcomb, Simon, *The Reminiscences of an Astronomer*, Boston and New York: Houghton, Mifflin, 1903, pp. 182–94, on p. 184.

41. Osterbrock, Donald E., 'The rise and fall of Edward S. Holden', *Journal for the History of Astronomy*, **15** (pt 2), June 1984, pp. 81–127, and pt 3, October 1984, pp. 151–76. See also Osterbrock (Ref. 2), pp. 233–40.

42. Osterbrock (Ref. 2), p. 72.

43. Asaph Hall to E. C. Pickering, 30 October 1877, quoted by Jones and Boyd (Ref. 5), p. 305.

44. Stone, Remington P. S., 'The Crossley reflector: a centennial review', *Sky and Telescope*, **58** (4), October 1979, pp. 307–11; **58** (5), November 1979, pp. 396–400.

45. Osterbrock (Ref. 41), pp. 155–71.

46. Newcomb (Ref. 40), pp. 192–3.

47. Keeler, James Edward, *Photographs of Nebulae and Clusters made with the Crossley Reflector, Publications of the Lick Observatory*, vol. VIII, 1908, on p. 11. The text of the Keeler memorial volume, pp. 11–29, was reprinted from the *Astrophysical Journal*, **11**, 1900, pp. 325 ff.

48. Osterbrock (Ref. 2), p. 347. The prime disadvantage of a parabolic reflector is that off-axis star images become fan-shaped, the more so, the closer one gets to the edge of the field. This is known as *coma*. This problem is eliminated in Schmidt, Maksutov, or Ritchey–Chrétien optical systems.

49. Stone, Remington P. S., 'Lick Observatory's Chile station', *Sky and Telescope*, **63** (5), May 1982, pp. 446–8.

50. Moore, J. H., 'Fifty years of research at the Lick Observatory', *Publications of the Astronomical Society of the Pacific*, **50**, August 1938, pp. 189–203, on p. 195.

51. Shane, C. D., 'Distribution of galaxies', in *Galaxies and the Universe*, eds A. Sandage, M. Sandage & J. Kristian, Chicago and London: University of Chicago Press, 1975, pp. 647–663. The Shane and Wirtanen counts are given as the number of galaxies in a square 10′ on a side. Their work was originally published in the *Publications of the Lick Observatory*, **22** (pt 1), 1967.

52. Robinson, L. B., and Wampler, E. J., 'The Lick Observatory image-dissector scanner', *Publications of the Astronomical Society of the Pacific*, **84**, February 1972, pp. 161–6.

53. Osterbrock, Donald E., 'Graduate astronomy education in the early days of Lick Observatory', *Mercury*, **9** (6), November–December 1980, pp. 151–6.

54. Moore, Ref. 50, p. 198.

55. Hale, George E., 'The Yerkes Observatory of the University of Chicago. I. Selection of the site', *Astrophysical Journal*, **5**, 1897, pp. 164–180, on p. 171.

56. Osterbrock, Donald E., 'First world astronomy meeting in America', *Sky and Telescope*, **56** (3), September 1978, pp. 180–3.

57. Quoted by Wright (Ref. 1), p. 131.

58. Wright (Ref. 1), p. 132.

59. Keeler's address delivered at the Yerkes dedication was published in the *Astrophysical Journal*, **6**, 1897, pp. 271–88.

60. Ibid., p. 275.

61. Berendzen, Richard, 'Origins of the American Astronomical Society', *Physics Today*, December 1974, pp. 32–9.

62. Hale (Ref. 55), pp. 169–70.

63. Wright (Ref. 1), p. 129.

64. Hale, George E., 'The aim of the Yerkes Observatory', *Astrophysical Journal*, **6**, 1897, pp. 310–21, on p. 317.

65. The reader will recall that Antonia Maury's spectral classification scheme included subclasses a, b, and c to denote a sequence of decreasing line widths. The stars of subclass c turned out to be giant stars, and they rotate more slowly than dwarf stars. The modern spectral classification scheme includes a spectral type (OBAFGKM) and a luminosity class (I = supergiants, III = giants, V = main sequence stars, VII = white dwarfs). From the spectrum it is possible to assign a spectral type and estimate the intrinsic luminosity of the star by comparing the spectrum to the spectra of certain 'standard stars'. See Morgan, W. W., and Keenan, P. C., 'Spectral classification', *Annual Review of Astronomy and Astrophysics*, **11**, 1973, pp. 29–50. They state (pp. 30–1):

    The MK system is a phenomenology of spectral lines, blends, and bands, based on a general progression of color index (abscissa) and luminosity (ordinate). It is defined by an array of standard stars, located on the two-dimensional spectral type vs luminosity-class diagram. These standard reference points do not depend on values of any specific line intensities or ratios of intensities; they have come to be defined by the appearance of the totality of lines, blends, and bands in the ordinary photographic region ... In the final analysis there is only one meaningful approach we can adopt in the empirical system: to *define it in terms of real objects, without comment.*

    This avoids the problem of redefining the spectral types of objects when our ideas of stellar evolution change.

    Regarding the UBV system of photoelectric photometry, see Johnson, Harold L. and Morgan, William W., 'Fundamental stellar photometry for standards of spectral type in the Revised System of the Yerkes Spectral Atlas', *Astrophysical Journal*, **117**, 1953, pp. 313–52, which is also to be found in Lang and Gingerich (Ref. 24), pp. 45–9. The UBV system is the most widely used photometric system. The three filters are ultraviolet (U, wavelength 3650 Å), blue (B, 4400 Å), and visual (V, 5500 Å). The color index of a star is $B - V$ (the difference of the blue and visual magnitudes). A0 stars have $B - V = 0.00$; cooler stars have positive color index; hotter stars have negative color index. A list of *UBV* standard star magnitudes and colors of 53 845 stars is to be found in *Astronomy and Astrophysics Supplement*, **34**, 1978, p. 1.

66. Otto Struve (1897–1963), the great-grandson of F. G. W. Struve, came to the United States in 1921. He had fought against the Bolsheviks during the Russian civil war. As a result of the efforts of E. B. Frost, the director of Yerkes Observatory, Struve was able to leave behind a miserable refugee existence in Istanbul. He earned his PhD from the University of Chicago in 1923. He became one of the foremost and most formidable astrophysicists of the twentieth century. A couple of examples will suffice concerning his personality. On the first day of Struve's tenure as director at Yerkes, one of the staff members was finishing up the reduction of a spectrogram, a task which the outgoing director had assigned to him. Struve asked the staff member what he was doing. 'Reducing this spectrogram'. 'Who told you to do that?' 'Dr Frost'. Struve's final comment was, 'Well, now things are going to be different'.

An astronomer I once talked to at the Lick observatory dining hall at Mt Hamilton describes a first meeting with Otto Struve something like this:

I went into his office, and there he was, the famous Otto Struve. He was about *eight* feet tall and had red hair. And he would look at you with one eye while staring out the window or scanning his book shelves with the other. Then, addressing me, he said, 'Vell, Mistah . . . , vot can I do forrr you today, hmmm?' I couldn't remember what I went to see him about.

The great number of publications authored by Otto Struve was due to his own methods, which included bringing along a tape recorder for dictation while he was observing to obtain data for some other project.

Chapter 7

1. Jones, Bessie Zaban and Boyd, Lyle Gifford, *The Harvard College Observatory. The First Four Directorships, 1839–1919*, Cambridge, Massachusetts: Belknap Press, 1971, pp. 256–68.
2. Wright, Helen, *Explorer of the Universe: a Biography of George Ellery Hale*, New York: E. P. Dutton, 1966, pp. 159–96.
3. Adams, Walter S., 'Early days at Mount Wilson', *Publications of the Astronomical Society of the Pacific*, **59**, 1947, pp. 213–31, 285–304.
4. Hale, George E., 'A study of the conditions for solar research at Mount Wilson, California', *Contributions from the Solar Observatory, Mt Wilson, California*, No. 1, [1903], p. 1. This also appeared in the *Astrophysical Journal*, March 1905.
5. Ibid., pp. 2–3.
6. Ibid., p. 27.
7. Ibid., p. 23.
8. Van Helden, Albert, 'Building large telescopes, 1900–1950', in *GHA 4A*, pp. 134–52, on p. 139.
9. Hale, George Ellery, 'On the probable existence of a magnetic field in Sun-spots', in Lang and Gingerich, pp. 96–103.

10. Hale, George E., 'The solar observatory of the Carnegie Institution of Washington', *Contributions from the Solar Observatory, Mt. Wilson, California*, No. 2, 1905, p. 3.
11. Wright (Ref. 2), pp. 227–31.
12. Van Helden (Ref. 8), p. 140. See also Osterbrock, Donald E., 'The quest for more photons: how reflectors supplanted refractors as the monster telescopes of the future at the end of the last century', *Astronomy Quarterly*, **5**, 1985, 87–95.
13. Lang and Gingerich, (Ref. 9), pp. 698–703.
14. Ibid., pp. 523–34.
15. Ibid., pp. 430–2.
16. Ibid., pp. 221–4.
17. Ibid., pp. 67–74.
18. Ibid., pp. 663–5.
19. Baliunas, Sallie L. *et al.*, 'Time-series measurements of chromospheric CaII H and K emission in cool stars and the search for differential rotation', *Astrophysical Journal*, **294**, 1 July 1985, pp. 310–25.
20. Adams (Ref. 3), pp. 214–15.
21. Ibid., p. 225.
22. Hale (Ref. 10), pp. 17–18.
23. Wright (Ref. 2), pp. 252–65, 318–31.
24. Ibid., p. 321. Adams incorrectly gives it as November 1st.
25. Adams (Ref. 3), p. 301.
26. Lang and Gingerich (Ref. 9), pp. 2–7.
27. King, Henry C., *The History of the Telescope*, New York: Dover, 1979, pp. 320–45, in particular pp. 337–8.
28. Lang and Gingerich, pp. 713–15.
29. Ibid., pp. 725–8.
30. Sandage, Allan and Tammann, G. A., 'Steps toward the Hubble constant. VII. Distances to NGC 2403, M 101, and the Virgo cluster using 21 centimeter line widths compared with optical methods: The global value of H_o', *Astrophysical Journal*, **210**, 1976, pp. 7–24.
31. Aaronson, Marc *et al.*, 'A distance scale from the infrared magnitude/HI velocity-width relation. III. The expansion rate outside the local supercluster', *Astrophysical Journal*, **239**, 1980, pp. 12–37. This whole method is based on the 'infrared Tully-Fisher relation', which has its roots in a paper by R. Brent Tully and J. Richard Fisher, 'A new method of determining distances to galaxies', *Astronomy and Astrophysics*, **54**, 1977, pp. 661–73.
32. For a review of the evolution of the Hubble constant see Gerard de Vaucouleurs, 'The distance scale of the universe', *Sky and Telescope*, **66** (6), December 1983, pp. 511–16.
33. Sandage, Allan, 'On the age of M 92 and M 15', *Astronomical Journal* **88** (8), August 1983, pp. 1159–65.
34. See Tinsley, Beatrice M., 'The cosmological constant and cosmological change', *Physics Today*, June 1977, pp. 32–8.
35. Islam, Jamal N., 'The ultimate fate of the universe', *Sky and Telescope*, **57** (1), January 1979, pp. 13–18.

36. Lang and Gingerich, pp. 744–52.
37. Irwin, John B., 'Chile's mountain observatories revisited', *Sky and Telescope*, **47** (1), January 1974, pp. 11–16.
38. Soderblom, David R. and Bond, Howard E., 'Obsolescence in telescopes?' *Physics Today*, April 1985, pp. 9, 11.
39. Quoted in Wright (Ref. 2), p. 417.
40. King (Ref. 27), p. 403.
41. Bowen, I. S., 'The 200-inch Hale telescope', in *Telescopes*, eds Gerard P. Kuiper and Barbara M. Middlehurst, Chicago and London: University of Chicago Press, pp. 1–15.
42. Hale, George Ellery, 'The 200-inch reflector on Mount Palomar', in *Source Book in Astronomy 1900–1950*, ed. Harlow Shapley, Cambridge, Massachusetts: Harvard University Press, 1960, pp. 3–12; Lang and Gingerich, pp. 21–6; King (Ref. 27), pp. 403–15; Van Helden (Ref. 8), pp. 144–52; and Wright (Ref. 2), pp. 387–408, 415–29.
43. I saw 55 gallon drums of such oil in the basement of Palomar Observatory when I visited there in August of 1984.
44. Lang and Gingerich (Ref. 9), pp. 803–18.
45. 'Breaking the redshift-4 barrier', *Science News*, **132**, 17 October 1987, p. 254.
46. For velocities (v) small compared to the speed of light (c), the formula for redshift is simply $\Delta\lambda/\lambda = Z = v/c$. The relativistic (high velocity) relation is

$$Z = \sqrt{\frac{1 + v/c}{1 - v/c}} - 1,$$

47. Oke, J. B., 'Palomar's Hale telescope: the first 30 years', *Sky and Telescope*, **58** (6), December 1979, pp. 505–9.
48. Donald E. Osterbrock, who was one of the first Caltech astronomy faculty members, has written an interesting account of the early days at Palomar in his biographical essay 'Rudolph Leo Minkowski 1895–1976', *Biographical Memoirs of the National Academy of Sciences*, **54**, 1983, pp. 270–98.
49. di Cicco, Dennis, 'Comet Halley found', *Sky and Telescope*, **64** (6), December 1982, p. 551.
50. Ioannisiani, Bagrat K., 'The Soviet 6-meter altazimuth reflector', *Sky and Telescope*, **54** (5), November 1977, pp. 356–62.

Chapter 8

1. Ronan, Colin A., 'Edmond Halley', in *DSB*, vol. 6, pp. 67–72; Ronan, Colin A., *Edmond Halley: Genius in Eclipse*, New York: Doubleday, 1969, 251 p.; a review of this book by P. L. Brown is in *Sky and Telescope*, **39** (1), January 1969, p. 41; see also Ashbrook, Joseph, 'Edmond Halley at St Helena', *Sky and Telescope*, **40** (2), August 1970, pp. 86–7.
2. Evans, David S., 'Nicolas de la Caille and the southern sky', *Sky and Telescope*, **60** (1), July 1980, pp. 4–7.

3. Stoy, R. H., 'Astronomy in South Africa', in *A History of Scientific Endeavour in South Africa*, ed. A. C. Brown, Cape Town: Royal Society of South Africa, 1977, pp. 409–26.
4. Ashbrook, Joseph, 'John Herschel's expedition to South Africa', *Sky and Telescope*, **37** (6), June 1969, pp. 58, 66. See also Evans, David S., 'John Frederick William Herschel', in *DSB*, vol. 6, pp. 323–8.
5. Forbes, Eric G., *Greenwich Observatory . . .* , London: Taylor & Francis, 1975, vol. 1, pp. 164–6.
6. Warner, Brian, *Astronomers at the Royal Observatory, Cape of Good Hope: a History with Emphasis on the Nineteenth Century*, Cape Town and Rotterdam: A. A. Balkema, 1979, 132 p.
7. Stoy, R. H., 'Astronomy', in *Standard Encyclopaedia of Southern Africa*, Cape Town: Nasou, 1970, vol. 1, pp. 583–9.
8. Anon., 'South African Astronomical Observatory/Suid-Afrikaanse Astronomiese Observatorium', published by the observatory in [1972]. In English and Afrikaans. 19 p.
9. Series of articles on 'Astronomy in the Republic/Sterrekunde in die Republiek', *Scientiae*, **9** (10), October 1968, pp. 2–20. In English and Afrikaans.
10. Evans, David S., 'Astronomical institutions in the southern hemisphere, 1850–1950', in *GHA 4A*, pp. 153–65.
11. Glass, I. S. (ed.), *South African Astronomical Observatory Facilities Manual 1982*, 3rd edn, Cape Town: SAAO.
12. By 'results per research dollar' I mean specifically the number of citations generated to published research which was carried out on telescopes of different sizes. See Abt, Helmut A., 'The cost-effectiveness in terms of publications and citations of various optical telescopes at the Kitt Peak National Observatory', *Publications of the Astronomical Society of the Pacific*, **92** (547), June 1980, pp. 249–54.
13. Herrmann, D. B., 'Zur Statistik von Sternwartengründungen im 19. Jahrhundert', *Die Sterne*, 49th year, no. 1, 1973, pp. 48–52; Herrmann, D. B., 'Sternwartengründungen, Wissensproduktion und ökonomischer Fortschritt', *Die Sterne*, 51st year, no. 4, 1975, pp. 228–34.
14. Price, Derek J. de Solla, *Little Science, Big Science*, New York and London: Columbia Univ. Press, 1963, p. 7.
15. Vickery, B. C., 'Bradford's law of scattering', *Journal of Documentation*, **4** (3), December 1948, pp. 198–203. If we rank scientific journals by the number of papers published in a particular field of research, each field will have a core group of journals which are those that are ranked at the top of such a list. Considering the top n journals (ranked by the number of papers published in a given field), if these generated P published papers, then it would take Qn of the next ranked journals to generate another P papers, and Q^2n of the still lower ranked journals to generate another P papers, with $Q \gg 1$.
16. Abt, Helmut A., 'Citations to single and multiauthored papers', *Publications of the Astronomical Society of the Pacific*, **96** (583), September 1984, pp. 746–9. For the 1969–1973 cumulative index of *Astronomy and Astrophysics Abstracts*, the number

of authors divided by the number of items referenced is 1.73. For the 1974–78 cumulative volume there is an average of 1.93 authors per published item. For the 1979–83 cumulative volume there is an average of 2.03 authors per published item.

17. See Price (Ref. 14), pp. 40–9. The power law distribution of scientific productivity was first discovered by Alfred J. Lotka, 'The frequency distribution of scientific productivity', *Journal of the Washington Academy of Sciences*, **16** (12), 1926, pp. 317–23. Lotka considered the number of chemists listed in *Chemical Abstracts*, who had published 1, 2, 3, ... papers during the years 1907–16, finding a power law index $\alpha = 1.888$. For the lifetime publication totals of physicists for the entire range of history up to 1900, as listed in Auerbach's *Geschichtstafeln der Physik*, he found a power law with index $\alpha = 2.021$. The numbers of papers presented by mathematicians at the Chicago meetings of the American Mathematical Society during the years 1897–1922 (Dresden, Arnold, 'A report on the scientific work of the Chicago Section, 1897–1922', *Bulletin of the American Mathematical Society*, **28**, 1922, pp. 303–7) and the articles in the first twenty volumes of *Econometrica*, 1933–52 (Leavens, Dickson H., *Econometrica*, **21**, 1953, pp. 630–2) both show the same statistical trend, with power law indices $\alpha = 1.5$ and 1.9, respectively. The most prolific astronomer on the basis of the number of published items was Ernst Öpik (1893–1985), with at least 1094 scientific items. 'This figure comprises 273 scientific papers, 514 individual articles from the News & Comments section of *The Irish Astronomical Journal*, 173 book reviews, 65 obituaries, 4 books and 65 other notes' [and 16 piano compositions]. (Letter of John McFarland, Librarian at Armagh Observatory, to K. Krisciunas, 27 February 1986.) Öpik's total is a far cry from the all-time record for scientific productivity, which is held by one T. D. A. Cockerell (1866–1948), a University of Colorado entomologist, who published the incredible sum of 3904 papers over a span of more than 50 years (or better than a paper a week for his entire career)! See Price, Derek J. de Solla, 'A general theory of bibliometric and other cumulative advantage processes', *Journal of the American Society of Information Science*, **27** (5), September–October 1976, pp. 292–306, on p. 300.

18. Price (Ref. 14), p. 45.

19. Markowitz, W., 'Polar motion: history and recent results', *Sky and Telescope*, **52** (2), August 1976, pp. 99–103, 108.

20. Newcomb, Simon, 'Aspects of American astronomy', *Astrophysical Journal*, **6**, 1897, pp. 289–309, on p. 303.

21. van de Hulst, H. C., 'Nanohertz astronomy', in *The early years of radio astronomy*, Ref. 94 below, pp. 385–98, in partic. pp. 386–9. See also Dadaev, A. N., *Pulkovo Observatory*, Springfield, Virginia: National Technical Information Service, 1978, pp. 55–6; and *DSB* article on Tikhov.

22. *Sosie* – (pronounced so–zy) applies to individual (galaxies), not closely related genetically, which are so much 'look-alikes' that they can act as doubles. (L. Bottinelli *et al.*, *Astrophysical Journal Suppl.*, **59**, 1985, p. 213). *Blazar* – BL Lacertae objects and other optically violently variable (OVV) objects (R. R. J. Antonucci and J. S. Ulvestad, *Astrophysical Journal*, **294**, 1985, p. 158). *Liner* – low-ionization nuclear emission-line region (T. M. Heckman, *Astronomy and Astrophysics*, **87**, 1980, p. 152). *Noisar* – X-ray binaries showing excess optical variability on time

scales of seconds (J. Middleditch, *Bulletin of the American Astronomical Society*, **13**, 1981, p. 816).

23. I principally rely here on Leo Goldberg's article, 'The founding of Kitt Peak', *Sky and Telescope*, **65** (3), March 1983, pp. 228–32; and:

24. Kloeppel, James E., *Realm of the Long Eyes: a Brief History of Kitt Peak National Observatory*, San Diego, California: Univelt, 1983, 136 p. I also rely on some KPNO brochures obtained January 1985.

25. Edmondson, Frank K., review of ibid., *Journal for the History of Astronomy*, **15** (2), June 1984, pp. 139–41.

26. Anon., 'Kitt Peak's 80-inch stellar telescope', *Sky and Telescope*, **23** (1), January 1962, pp. 5–9. It was called an 80 inch because the specifications called for an 80 inch mirror that could concentrate 75% of the light within an image 0.3 arc seconds in diameter at the Cassegrain focus.

27. McMath, Robert R. and Pierce, A. Keith, 'The large solar telescope at Kitt Peak', *Sky and Telescope*, **20** (2), August 1960, pp. 64–7, and **20** (3), September 1960, pp. 132–5.

28. See *Sky and Telescope*, **73** (4), April 1987, p. 379.

29. Hall, D. N. B., Kleinmann, S. G. and Scoville, N. Z., 'Broad helium in the Galactic Center', *Astrophysical Journal*, **260**, 1982, pp. L53–L57.

30. Aaronson, Marc *et al.*, *Astrophysical Journal*, **239**, 1980, pp. 12–37, and references therein. Much of the infrared data was obtained with the Kitt Peak No. 3 16 inch, the No. 1 36 inch, and the 84 inch.

31. Walsh, D., Carswell, R. F. and Weymann, R. J., '0957 + 561 A, B: twin quasistellar objects or gravitational lens?' *Nature*, **279**, 31 May 1979, pp. 381–4. This was the first example of the gravitational lens effect. The observations were made on the Kitt Peak 84 inch with an image dissector scanner.

32. McCoy, Patrick C., 'Cultural, historical view of Mauna Kea', *Hawaii Tribune-Herald*, January 27, 1980, p. B-3.

33. *Missionary Album: Portraits and Biographical Sketches of the American Protestant Missionaries to the Hawaiian Islands*, Honolulu: Hawaiian Mission Children's Society, 1969, pp. 102–3. Goodrich lived from 1794 to 1852. He was in Hawaii from 1823 to 1836.

34. Ellis, William, *A Narrative of a Tour Through Hawaii, or Owhyhee; with Remarks on the History, Traditions, Manners, Customs and Language of the Inhabitants of the Sandwich Islands*, Honolulu: Hawaiian Gazette Co., 1917, pp. 4, 301–4.

35. Goodrich, Joseph, 'Volcanic character of the Island of Hawaii', *American Journal of Science and Arts*, **11**, October 1826, pp. 2–7, on p. 5.

36. Byron, [George A.], *Voyage of H. M. S. Blonde to the Sandwich Islands, in the Years 1824–1825*, London: John Murray, 1826, pp. 169–73. Macrae and Goodrich are not specifically named, but Macrae is named in other articles, and I have little doubt that the 'missionary' who accompanied Macrae was Goodrich.

37. Goodrich, Joseph, 'Notices of some of the volcanos and volcanic phenomena of Hawaii, (Owyhee,) and other islands in that group', *American Journal of Science and Arts*, **25**, January 1834, pp. 199–203, on p. 200.

38. [Hooker, W. J., compiler], 'A brief memoir of the life of Mr David Douglas, with extracts from his letters', *Hawaiian Spectator*, **2**, 1839, pp. 396–437, on p. 405.

39. Jarves, James J., *Scenes and Scenery in the Sandwich Islands, and a Trip through Central America: being Observations from my Note-book during the Years 1837–1842*, Boston: James Munroe, 1843, pp. 224–32.

40. Wilkes, Charles, *Narrative of the United States Exploring Expedition during the Years 1838, 1839, 1840, 1841, 1842*, Philadelphia: Lea & Blanchard, 1845, vol. 4, pp. 200–3. The three who made it to the summit of Mauna Kea were Dr Charles Pickering (a naturalist), [J. D.] Brackenridge (assistant botanist), and a man named Dawson.

41. Wentworth, Chester K., 'Mauna Kea, the White Mountain of Hawaii', *Mid-Pacific Magazine*, October-December 1935, pp. 290–6, on p. 292. Wentworth led the expedition of August 1935, which was sponsored by the Hawaiian Academy of Science. See also Bryan, E. H., Jr (ed.) *Mauna Kea Here we Come: the Inside Story of a Scientific Expedition*, Honolulu: Bryan, 1979, 78 p., which also contains a bibliography of non-astronomical articles about Mauna Kea.

42. Forster, Peter J. G., 'Health and work at high altitude: a study at the Mauna Kea Observatories', *Publications of the Astronomical Society of the Pacific*, 96 (580), June 1984, pp. 478–87.

43. Letter of Thomas H. Hamilton to Mitsuo Akiyama, 15 July 1963.

44. Anonymous, 'Unsung heroes', *Hawaii Business*, November 1979, pp. 53 ff.; also Akiyama, Mitsuo, compiler, 'Chronology of Mauna Kea summit road and observatory station', [20 July 1964], hereafter referred to as Akiyama Chronology. Akiyama's personal archives contain copies of much of the correspondence relating to the establishment of the first telescope at Mauna Kea.

45. GPK (Gerard P. Kuiper) to Fujio Matsuda, 18 March 1964.

46. Ibid.

47. Ibid. According to the University of Arizona proposal to build a 60 inch telescope at Mauna Kea (dated December 1964), which summarizes the initial site tests, some sky tests had been carried out at Mauna Loa Observatory in June of 1963 with a 6 inch telescope.

48. GPK to Governor John A. Burns, 17 January 1964; GPK to Akiyama, 23 January 1964, which quotes Kuiper's report to NASA Headquarters of 22 January.

49. Goodrich Cone and Goodrich Pass (where the present road passes between the summit cone and Goodrich Cone) are obviously so named in honor of Joseph Goodrich, the first non-Hawaiian to visit the top. The official name of Goodrich Cone is now Pu'u Haukea which means 'frosty white cinder cone'. In 1974 the old informal names for some of the summit cones were changed to Hawaiian names by the Hawaii Committee on Geographical names. Macrae Cone to the north of the summit cone became Pu'u Hao'oki; Douglas Cone to the northwest of Pu'u Poli'ahu became Pu'u Pohaku; and the summit cone was named Pu'u Wekiu, meaning 'summit cone'. See Cruikshank, Dale P., *Mauna Kea: a Guide to the Upper Slopes and Observatories*, Honolulu: University of Hawaii, Institute for Astronomy, 1986, pp. 24–6.

50. Lyman Nichols to GPK, 12 February 1964; GPK to Matsuda, 18 March 1964.

51. Akiyama notes; *Hawaii Tribune-Herald*, 26 March 1964; Akiyama Chronology.

52. Akiyama Chronology.

53. Burns to Akiyama, 10 March 1964.

54. Telephone conversation with Carlton M. Gillespie, Jr, 22 February 1985. Gillespie

worked for Kuiper and supervised the construction of the $12\frac{1}{2}$ foot dome on Mauna Kea in 1964. He became involved with Kuiper's efforts to carry out astronomy from airplanes in the late 1960s and has flown hundreds of flights as a Mission Director on the Kuiper Airborne Observatory.

55. Akiyama Chronology; *Hawaii Tribune-Herald*, 13 April 1964.
56. Akiyama Chronology.
57. Telephone conversation with Gillespie, 22 February 1985.
58. Akiyama Chronology.
59. In a letter of 11 March 1985 (AKH to K. Krisciunas) Herring relates:

> You ask how good is the mirror? Frankly, I don't know. The figure
> has failed to break down under any test that I have been able to
> devise for it. I do know that the mirror gives exquisite performance
> and surpasses anything else of equal size that I have ever seen. During
> [the years I made telescope optics] I made over 3500 paraboloids, but
> I don't think I ever topped this mirror. It is truly one of a kind. I have
> held on to it in spite of some very generous offers and one attempted
> theft.

> Gillespie has echoed these sentiments. In a letter of 30 December 1985 (AKH to
> KK) Herring adds: 'I am not now using the supermirror [.] I was always afraid
> that something might happen to it so about 10 years ago I made a replacement
> and retired the jewel. It has been resting in its box in a corner of my study for lo
> these many moons'.

60. AKH to KK, 11 March 1985.
61. *Hawaii Tribune-Herald*, 13 June 1964.
62. *Honolulu Star Bulletin*, 21 July 1964; conversation with Akiyama, 15 March 1985. Akiyama tells me that the ceremony was to have taken place at Pu'u Poli'ahu, but it was too windy. Some people stayed at the mountain that day to look through the telescope.
63. Kuiper, Gerard P., 'Address given at Mauna Kea Station Dedication', 20 July 1964. From Akiyama's archives. Akiyama, Herring and Gillespie all remind me of the major role played by William Seymour of Hilo in the establishment of the first Mauna Kea station. Seymour (1898–   ), a ham radio operator, took care of the radio communications back and forth to the mountain and back and forth to Tucson.
64. Johnson, H. L and Kuiper, G. P., 'Proposal to National Aeronautics and Space Administration for the design, construction, and installation of a 60-inch telescope on Mauna Kea, Hawaii in collaboration with the University and the State of Hawaii', December 1964, p. 5.
65. Walker, Merle F., 'How good is your observing site?' *Sky and Telescope* **71** (2), February 1986, pp. 139–43.
66. AKH to KK, 11 March 1985; telephone conversation with Hartmann, 15 March 1985. Hartmann only took part in the first observing run, having arrived on 24 June (according to Akiyama's Chronology). There was only a 1 or 2 day overlap between Herring and Hartmann. When Herring was in Hawaii the telescope had his supermirror, but Hartmann used the mirror made by the LPL optical shop. Hartmann stayed through late July.

67. Gillespie to Akiyama, 30 September 1964; *Hawaii Tribune-Herald*, 27 October 1964.

68. GPK to Woollard, 28 October 1964.

69. Memo of GPK to NASA of 17 November 1964; and letter of GPK to Homer Newell of NASA, 18 November 1964.

70. GPK to Woollard, 14 December 1964.

71. Woollard to Kuiper, 28 October 1964.

72. Woollard to Major General Edmond H. Leavey, Chairman of the Governor's Advisory Committee on Science and Technology, 5 January 1965.

73. See Ref. 64, above.

74. Jefferies, John T., principal investigator. 'A proposal to the National Aeronautics and Space Administration for the construction of an 84-inch astronomical telescope, coudé spectrograph, and associated equipment for lunar, planetary, and stellar observations in the Hawaiian Islands', University of Hawaii, Astrophysics and Atmospheric Physics Section, Hawaii Institute of Geophysics, February 1965.

75. GPK to Homer Newell, 30 March 1965, referring to a clipping from the *Honolulu Advertiser* of 26 March.

76. Jefferies relates:

> There were endless people telling me at that time (1965) that Mauna Kea was *far* too high for serious consideration as a site for a large telescope. That was the primary reason (along with strong political Island factors, the presence of the UH Solar Observatory, and existence of logistical support – still not in-place on Mauna Kea) that led to the inclusion of Haleakala in the survey. It became clear very quickly that M[auna] K[ea] was both superior and viable and its selection was announced about 6 months after site testing began.

> (JTJ to K. Krisciunas, 20 March 1985.) See also Davies, Lawrence E., 'Group making surveys for highest observatory', *Honolulu Star-Bulletin*, 29 June 1966, in which Jefferies is quoted as saying, 'The story of the site survey is a real epic. We had to start from scratch, accumulate a staff and beg or borrow money. We got some young college graduates, some amateur astronomers and a young Swiss boy who was on a trip around the world and asked for a job'.

77. Waldrop, M. Mitchell, 'Mauna Kea (I): Halfway to space', *Science*, **214**, 27 November 1981, pp. 1010–13; 'Mauna Kea (II): Coming of age', ibid., **214**, 4 December 1981, pp. 1110–14. According to Gillespie, Waldrop incorrectly states that Kuiper participated in the actual grading of the first road.

78. According to Ewen Whitaker of the Lunar and Planetary Laboratory. Phone conversation with KK, 7 March 1984.

79. Southward, Walt, '$3 million observatory started', *Honolulu Advertiser*, 22 September 1967; Jefferies, John T., and Sinton, William M., 'Progress at Mauna Kea Observatory', *Sky and Telescope*, **36** (3), September 1968, pp. 140–5; Anonymous, 'Mauna Kea Observatory dedicated', ibid., **40** (5), November 1970, pp. 276–7; Morrison, David, and Jefferies, John T., 'Hawaii's Mauna Kea Observatory today', ibid., **44** (6), December 1972, pp. 361–5. See also Waldrop (Ref. 77) above.

80. The first seven were the Hale 5 m (1948), the Lick 3 m (1959), the McDonald 107

inch (1968), the 2.6 m Shajn telescope of the Crimean Astrophysical Observatory (1961), the Mt Wilson 100 inch (1917), the Isaac Newton Telescope (98 inch; 1967), and the Steward Observatory 90 inch (1969) at Kitt Peak.

81. Cruikshank, Dale P., '20th-century astronomer', *Sky and Telescope*, **47** (3), March 1974, pp. 159–64.

82. Other mountains 'pioneered' by Kuiper include: (*a*) The observatories in Chile at Cerro Tololo, La Silla, and Las Campanas. Kuiper initiated the program of site testing in Chile in 1959 at the request of Dr. Rutllant, the Director of the Observatory of the University of Chile. (*b*) The Catalina Observatory of the Lunar and Planetary Laboratory and Mt Lemmon Infrared Observatory, northeast of Tucson. (*c*) The Smithsonian Observatory at Mt Hopkins. Kuiper did no site testing, but flew over the area extensively and recommended it to Fred Whipple from the general lack of turbulence and favorable location. (*d*) San Pedro Martir (N. Baja California, Mexico). Herring did the testing there for G. Haro of the University of Mexico. (Letter of Ewen Whitaker to K. Krisciunas, 14 March 1985)

According to Whitaker, Kuiper examined other mountain sites but did not pursue them for one reason or another. These were: (*a*) Mt Graham, near Safford, Arizona; (*b*) San Francisco Peaks, Arizona; (*c*) White Mountain, on the Nevada-California border; (*d*) Mt Shasta, California; (*e*) Pikes Peak, Colorado; (*f*) Cone Peak, California; (*g*) Junipero Serra Peak, California; (*h*) the southern tip of Baja California.

83. Hastings, Barbara, 'The house that John built', *Honolulu Advertiser*, 15 July 1983, section D2, p. 1. See also Waldrop (Ref. 77). Jefferies, originally from Australia and educated at Cambridge, came to Hawaii in 1964 from the Joint Institute for Laboratory Astrophysics in Boulder, Colorado. He was the first Director of the University of Hawaii's Institute for Astronomy when it was formally established on 1 July 1967. In 1983 he left to become the first 'superdirector' of Kitt Peak, Sacramento Peak, and Cerro Tololo Observatories, now called the National Optical Astronomy Observatories (NOAO), a post he held until March 1987.

84. Sanchez, F., 'Astronomy in the Canary Islands', *Vistas in Astronomy*, **28**, 1985, pp. 417–30, on p. 427.

85. Struve, Otto Wilhelm, *Erinnerung an den Vater*, Karlsruhe, 1896. I only have an unpublished translation of this by Alan H. Batten. The particular information about Wilhelm Struve's first wedding is on p. 28 of the 87 page translation.

86. Wright, Helen, *Explorer of the Universe*, New York: E. P. Dutton, 1966, p. 71.

87. Osterbrock, Donald E., 'The rise and fall of Edward S. Holden: part 2', *Journal for the History of Astronomy*, **15** (3), October 1984, pp. 151–76, on p. 151. A history of the Pic du Midi Observatory can be found in: Rösch, Jean and Dragesco, Jean, 'The French quest for high resolution', *Sky and Telescope*, **59** (1), January 1980, pp. 6–13.

88. Sanchez (Ref. 84), on p. 421.

89. Henbest, Nigel, 'British astronomy reaches new heights on La Palma', *Astronomy*, **13** (4), April 1985, pp. 6–22; Penston, Michael, 'INT first light', *Gemini* (Newsletter of the Royal Greenwich Observatory), no. 10, April 1984, pp. 1–2.

90. Smith, F. Graham and Dudley, J., 'The Isaac Newton Telescope', *Journal for the History of Astronomy*, **13** (1), 1982, pp. 1–18; Laing, R. and Jones, D., 'The Isaac Newton Group', *Vistas in Astronomy*, **28**, 1985, pp. 483–503.

91. Boksenberg, A., 'The William Herschel Telescope', *Vistas in Astronomy*, **28**, 1985, pp. 531–53.

92. Sullivan, Woodruff T., III, 'Radio astronomy's golden anniversary', *Sky and Telescope*, **64** (6), December 1982, pp. 544–50.

93. Kellermann, K. and Sheets, B. (eds), *Serendipitous Discoveries in Radio Astronomy*, Green Bank, West Virginia: National Radio Astronomy Observatory, 1983, hereafter referred to as KS.

94. Sullivan, W. T., III, (ed.), *The Early Years of Radio Astronomy: Reflections Fifty Years After Jansky's Discovery*, Cambridge University Press, 1984, hereafter referred to as Sullivan.

95. Reber, G., 'Early radio astronomy at Wheaton, Illinois', in Sullivan, ibid., pp. 43–66. See also Reber, Grote, 'Radio astronomy between Jansky and Reber', in KS (Ref. 93), pp. 71–8.

96. Reber, G. and Greenstein, J. L., 'Radio frequency observations of astronomical interest', *Observatory*, **67**, 1947, pp. 15–26.

97. Greenstein, Jesse L., 'Optical and radio astronomers in the early years', in Sullivan (Ref. 94), pp. 67–81, on pp. 71–2.

98. See Bowen, E. G., 'The origins of radio astronomy in Australia', in Sullivan (Ref. 94), pp. 84–111; Christiansen, W. N., 'The first decade of solar radio astronomy in Australia', ibid., pp. 112–31; and Kerr, F. J., 'Early days in radio and radar astronomy in Australia', ibid., pp. 132–45.

99. Smith, F. G., 'Early work on radio stars at Cambridge', in Sullivan (Ref. 94), pp. 236–48; Scheuer, P. A. G., 'The development of aperture synthesis at Cambridge', ibid., pp. 249–65; Lovell, Bernard, obituary of Martin Ryle, *Quarterly Journal of the Royal Astronomical Society*, **26**, 1985, pp. 358–68.

100. Lovell, Sir Bernard, 'The origins and early history of Jodrell Bank', in Sullivan (Ref. 94), pp. 192–211; Hanbury Brown, R., 'Paraboloids, galaxies and stars: memories of Jodrell Bank', ibid., pp. 212–35.

101. Bracewell, R. N., 'Early work on imaging theory in radio astronomy', in Sullivan (Ref. 94), pp. 166–90.

102. Mills, B. Y., 'Radio sources and the log *N*–log *S* controversy', in Sullivan (Ref. 94), pp. 146–65.

103. Bennett, A. S., 'The revised 3C catalogue of radio sources', *Memoires of the Royal Astronomical Society*, **58**, 1962, pp. 163–72.

104. Muller, C. A., 'Early galactic radio astronomy at Kootwijk', in *Oort and the Universe: a Sketch of Oort's Research and Person*, eds van Woerden, Hugo, Brouw, Willem N. and van de Hulst, Henk C., Dordrecht, Holland: D. Reidel, 1980, pp. 65–70; Christiansen, W. N., 'Oort and his large radiotelescope', ibid., pp. 71–8; Allen, R. J. and Ekers, R. D., 'Ten years of discovery with Oort's synthesis radio telescope', ibid., pp. 79–110; *Astronomy and Astrophysics* **31**, 1974, pp. 323–31.

105. Mayer, Cornell H., 'Early observations of thermal planetary radio emission', in KS (Ref. 93), pp. 266–74.

106. See Salomonivich, A. E., 'The first steps of Soviet radio astronomy', in Sullivan (Ref. 94), pp. 268–88; Ginzburg, V. L., 'Remarks on my work in radio astronomy', ibid., pp. 289–302; and Korolkov, D. V. and Pariiskii, Yu. N., 'The Soviet RATAN-600 radio telescope', *Sky and Telescope*, **57** (4), April 1979, pp. 324–9. The Soviets' efforts were ahead of everyone else as far as radar goes. Using an array of eight

16 m dishes located in the Crimea, they were the first to obtain radar signals reflected off the planets Venus, Mars and Mercury. Those observations of the early 1960s were carried out at a wavelength of 40 cm. The Radio Telescope of the Academy of Sciences (RATAN) is a ring of 895 panels 576 m in diameter. Its collecting area is 13 000 square meters (equal to that of the Very Large Array, or VLA). It is designed to work at the short end of the radio spectrum, from 8 mm to 20 cm.

107. Findlay, John W., 'The National Radio Astronomy Observatory', *Sky and Telescope*, **48** (6), December 1974, pp. 352–60. See also Anonymous, *National Radio Astronomy Observatory*, Green Bank, West Virginia: NRAO, [1968 booklet], 44 p.

108. Drake, Frank D., 'Discovery of the Jupiter radiation belts', in KS (Ref. 93), pp. 258–65, on pp. 263–4.

109. Shawcross, William E., 'Arecibo Observatory today', *Sky and Telescope*, **43** (4–5), April and May 1972, pp. 214–17, 228, 293–5; Anonymous, 'Arecibo's giant radio telescope upgraded', ibid., **49**, (3), March 1975, pp. 140–2, 146; Anonymous, 'Arecibo upgraded', ibid., **70**, (2), August 1985, p. 106.

110. Pettengill, Gordon, 'Discovery of Mercury's rotation', in KS (Ref. 93), pp. 275–9.

111. Bell Burnell, Jocelyn, 'The discovery of pulsars', in KS (Ref. 93), pp. 160–70.

112. Hachenberg, O., 'The new Bonn 100-meter radio telescope', *Sky and Telescope*, **40** (6), December 1970, pp. 338–43.

113. Heeschen, David S., 'The Very Large Array', *Sky and Telescope*, **49** (6), June 1975, pp. 344–51; Anonymous, 'The VLA takes shape', ibid., **52** (5) November 1976, pp. 320–2; Federer, C. A., Jr, 'The VLA: ears on the universe', ibid., **60** (6), December 1980, pp. 472–3. Examples of VLA maps can be found in Hjellming, Robert M., Bignell, R. Carl and Balick, Bruce, 'Mapping planetary nebulae with the VLA', ibid., **56** (3), September 1978, pp. 199–200; and in Righini-Cohen, R., Simon, M. and Felli, M., 'Unveiling interstellar clouds', ibid., **62** (3), September 1981, pp. 225–7.

114. Lo, K. Y., Backer, D. C., Ekers, R. D., Kellermann, K. I., Reid, M. and Moran, J. M., 'On the size of the galactic centre compact radio source: diameter < 20 AU', *Nature*, **315**, 9 May 1985, pp. 124–6.

115. Harwit, Martin, *Cosmic Discovery: the Search, Scope and Heritage of Astronomy*, New York: Basic Books, 1981, pp. 140–3, and references therein.

116. Penzias, Arno A. and Wilson, Robert W., 'A measurement of excess antenna temperature at 4080 MHz', *Astrophysical Journal*, **142**, 1965, pp. 419–21. See also Wilkinson, David T. and Peebles, P. J. E., 'Discovery of the 3 K radiation', in KS (Ref. 93), pp. 175–84; and Wilson, Robert W., 'Discovery of the cosmic microwave background', ibid., pp. 185–95.

117. Tanaka, Haruo, 'Development of solar radio astronomy in Japan up until 1960', in Sullivan (Ref. 94), pp. 335–48, on p. 341.

118. Covington, Arthur E., 'Beginnings of solar radio astronomy in Canada', in Sullivan (Ref. 94), pp. 317–34, on p. 329.

119. Harwit (Ref. 115), pp. 259–62.

120. See Gatland, Kenneth, principal author, *The Illustrated Encyclopedia of Space Technology: A Comprehensive History of Space Exploration*, New York: Salamander Books, 1981.

121. Low, Frank J., 'Airborne infrared astronomy: the early days', in *Airborne Astronomy*

Symposium, eds Harley A. Thronson and Edwin F. Erickson, Moffett Field, California: NASA Ames Research Center [1985], NASA Conference Publication 2353, pp. 1–8.

122. Russell, Ray W., 'The Lear Jet Observatory – fifteen years of discovery and rebirth', ibid., pp. 26–32.

123. Cameron, Robert M., 'NASA's 91-cm airborne telescope', *Sky and Telescope*, **52**(5), November 1976, pp. 327–31.

124. Larson, Harold P., 'Infrared spectroscopic observations of the outer planets, their satellites, and the asteroids', *Annual Review of Astronomy and Astrophysics*, **18**, 1980, pp. 43–75; Larson, Harold P., 'Exploration of the solar system by airborne astronomy', in *Airborne Astronomy Symposium* (Ref. 121), pp. 39–57.

125. Loewenstein, R. F., Harper, D. A., Hildebrand, R. H., Keene, Jocelyn, Orton, G. S. and Whitcomb, S. E., 'Far infrared and submillimeter observations of the giant planets', in *Airborne Astronomy Symposium* (Ref. 121), pp. 81–6.

126. This basic idea comes from Eric Becklin and Gerry Neugebauer when they were both at Caltech.

127. Krisciunas, K., 'Toward the resolution of the local missing mass problem', *Astronomical Journal*, **82**(3), March 1977, pp. 195–7; Bahcall, John N., 'K giants and the total amount of matter near the Sun', *Astrophysical Journal*, **287**, 1984, pp. 926–44, and references therein.

128. Soifer, B. T. and Pipher, Judith L., 'Instrumentation for infrared astronomy', *Annual Review of Astronomy and Astrophysics*, **16**, 1978, pp. 335–69, in particular pp. 343–9.

129. Robinson, Leif J., Schorn, Ronald A. and Beatty, J. Kelly, 'The frigid world of IRAS', *Sky and Telescope*, **67**(1–2), January and February 1984, pp. 4–8, 119–24.

130. Beichman, C. A., Neugebauer, G., Habing, H. J., Clegg, P. E. and Chester, T. J., eds., *Infrared Astronomical Satellite (IRAS). Catalogs and Atlases. Explanatory Supplement*, Pasadena, California: Joint IRAS Science Working Group, [preliminary version, 1984]. *Astronomy and Astrophysics Suppl.* **65**(4), September 1986.

131. Watts, Raymond N., Jr, 'The Celescope experiment', *Sky and Telescope*, **36**(4), October 1968, pp. 228–30; Anonymous, 'An observatory in space', ibid., **37**(1), January 1969, p. 17; Watts, Raymond N., Jr, 'Some early results from Celescope', ibid., **37**(5), May 1969, pp. 280–1.

132. Watts, Raymond N., Jr, 'An astronomy satellite named Copernicus', *Sky and Telescope*, **44**(4), October 1972, pp. 231–2, 235.

133. Underhill, Anne B., 'The International Ultraviolet Explorer satellite', *Sky and Telescope*, **46**(6), December 1973, pp. 377–9; Boggess, A., [and 32 coauthors], 'The IUE spacecraft and instrumentation', *Nature*, **275**(5679), 5 October 1978, pp. 372–7, and subsequent articles; Henbest, Nigel and Marten, Michael, *The New Astronomy*, Cambridge University Press, 1983, pp. 152–9.

134. Washburn, Mark, 'Maxwell's last frontier', *Sky and Telescope*, **69**(3), March 1985, p. 212.

135. Tucker, Wallace and Giacconi, Riccardo, 'The birth of X-ray astronomy', *Mercury*, **14**(6), November/December 1985, pp. 178–83, 190–1, and **15**(1), January/February 1986, pp. 13–18.

136. Henbest and Marten (Ref. 133), pp. 192–202; Giacconi, Riccardo, 'Progress in X-

ray astronomy', *Physics Today*, **26**(5), May 1973, pp. 38–47; Friedman, Herbert and Wood, Kent S., 'X-ray astronomy with HEAO 1', *Sky and Telescope*, **56**(6), December 1978, pp. 490–4.

137. Overbye, Dennis, 'The X-ray eyes of Einstein', *Sky and Telescope*, **57**(6), June 1979, pp. 527–34; Giacconi, R. [and 30 coauthors], 'The Einstein (HEAO 2) Observatory', *Astrophysical Journal*, **230**, 1979, pp. 540–50; Giacconi, Riccardo and Tananbaum, Harvey, 'The Einstein Observatory: new perspectives in astronomy', *Science*, **209**(4459), 22 August 1980, pp. 865–76.

138. Anonymous, 'New space observatories', *Sky and Telescope*, **66**(3), September 1983, pp. 215–16; Cordova, France A. and Mason, Keith O., 'Exosat: Europe's new X-ray satellite', ibid., **67**(5), May 1984, pp. 397–401. Anonymous, 'Exosat mission ends', in ibid., **72**(1), July 1986, p. 29.

139. Henbest and Marten (Ref. 133), pp. 202–5; Bignami, Giovanni F., 'Gamma-ray astronomy comes of age', *Sky and Telescope*, **70**(4), October 1985, pp. 301–4; Fichtel, Carl E. and Trombka, Jacob I., *Gamma Ray Astrophysics: New Insights into the Universe*, Washington, DC.: NASA, 1981, NASA SP-453, 401 p.

Chapter 9

1. Hoffmann, William F., 'The MMT program', in *The MMT and the Future of Ground-based Astronomy*, ed. Trevor C. Weeks, Smithsonian Astrophysical Observatory Special Report 385, 1979, pp. 23–35, referred to hereafter as Weeks.

2. Beckers, Jacques M., *et al.*, 'The Multiple Mirror Telescope', in *Telescopes for the 1980s*, eds G. Burbidge & A. Hewitt, Palo Alto, California: Annual Reviews, 1981, pp. 63–128.

3. Robinson, Leif J., 'Monster mirrors and telescopes', *Sky and Telescope*, **59**(6), June 1980, pp. 469–77.

4. Shectman, S. A., Hiltner, W. A., and Persson, S. E., *Bulletin of the American Astronomical Society*, **18**(4), 1986, p. 956.

5. Capriotti, E. R., and Baldwin, J. A., ibid., p. 955.

6. Enard, D., 'Very Large Telescope: recent developments', *The ESO Messenger*, No. 44, June 1986, pp. 37–40.

7. Osterbrock, Donald E., 'The future of ground-based astronomy I', in Weeks (Ref. 1), pp. 139–49; Sweet, William, 'Keck Foundation offers Caltech $70 million for 10-m telescope', *Physics Today*, February 1985, pp. 71–5.

8. Angel, J. R. P. and Woolf, N. J., '15 meter multiple mirror telescope design study', *Astrophysical Journal*, **301**, 1986, pp. 478–501; Field, George B. and Chaisson, Eric J., *The invisible universe: probing the frontiers of astrophysics*, Boston: Birkhäuser, 1985, p. 55 ff.

9. Basov, N. G., *et al.*, 'New astronomical telescope AST-1200 with a segmented, actively controlled primary mirror', in Weeks (Ref. 1), pp. 185–9; Steshenko, N. V., 'On the feasibility of the 25-meter optical telescope', in Weeks (Ref. 1), pp. 191–7; Robinson (Ref. 3).

10. O'Dell, C. R., 'The Space Telescope', in Burbidge and Hewitt (Ref. 2), pp. 129–94; Tucker, Wallace, 'The Space Telescope Science Institute', *Sky and Telescope*, **69**(4), April 1985, pp. 295–9, and following articles; van den Bergh, Sidney, 'Astronomy with the Hubble Space Telescope', *Journal of the Royal Astronomical Society of Canada*, **79**(3), 1985, pp. 134–42; *Sky and Telescope*, **73**(2), February 1987, p. 147.

11. Osterbrock (Ref. 6), p. 139.
12. Anonymous, 'NASA's SIRTF recast as orbiting free-flyer', *Sky and Telescope*, **68** (5), November 1984, pp. 412–3.
13. Field and Chaisson (Ref. 7), pp. 39 ff.
14. Smith, David H., 'Australia's bicentennial bonanza', *Sky and Telescope*, **65** (2), February 1983, pp. 120–1.
15. Gordon, Mark A., 'VLBA – a continent-size radio telescope', *Sky and Telescope*, **69** (6), June 1985, pp. 487–90; Broad, William J., 'Work begins on world's largest telescope', *New York Times*, December 3, 1985 (*Science Times* section), pp. 15, 17.

Index